# Laboratory Experiments USING MICROWAVE HEATING

NICHOLAS E. LEADBEATER

CYNTHIA B. McGOWAN

CRC Press
Taylor & Francis Group
Boca Raton London New York

CRC Press is an imprint of the
Taylor & Francis Group, an **informa** business

Cover art by Amanda Joy Heggland

CRC Press
Taylor & Francis Group
6000 Broken Sound Parkway NW, Suite 300
Boca Raton, FL 33487-2742



Printed on acid-free paper
Version Date: 20130311

International Standard Book Number-13: 978-1-4398-5609-3 (Paperback)



**Visit the Taylor & Francis Web site at**
**http://www.taylorandfrancis.com**

**and the CRC Press Web site at**
**http://www.crcpress.com**

# Contents

# Preface

We live in a fast-paced world, expecting fast communication and fast transportation. When we do a chemical reaction, we want it to be fast, too. For this, we can often turn to a technology that started out in the home kitchen: microwave energy. Arriving home hungry after a busy day and wanting a quick way to heat up something to eat, we of course turn to the microwave oven. Microwave energy has offered a fast and easy way to heat food for several decades. Interestingly, the Dow Chemical Company filed a patent in 1969 in which they documented carrying out chemical reactions using microwave energy but it was in 1986 that the first reports appeared in the scientific literature showing that microwave heating can be used in organic chemistry. Since those early days, the use of microwave heating as a tool in preparative chemistry has become mainstream in both industrial and academic settings. The development of specialized scientific microwave equipment has played a great part in this. Advantages over household microwave ovens include accurate measurement of parameters such as temperature and pressure as well as, most importantly, safety.

Microwave heating has also affected the undergraduate teaching laboratory. A standard organic chemistry laboratory manual covers a range of synthetic transformations. A number of these, as well as others, require extensive periods of heating to obtain high product yields. As a consequence students in the laboratory cannot perform these reactions because they do not fit into the time of an average laboratory period. Using microwave heating as a tool, these time constraints can often be overcome. Furthermore, if a reaction is complete within a few minutes of heating, students may be able either to repeat an experiment in the case of a mistake in preparing the reaction mixture or, more excitingly, to perform additional self-designed experiments based around the general reaction investigated in the lab period.

This book contains over 20 undergraduate laboratory experiments that can be performed using microwave heating. They range from organic to metal-catalyzed to inorganic and analytical chemistry in scope. A number of the transformations utilize water as a solvent, versus classical organic solvents, maintaining high yields and contributing to more environmentally friendly and sustainable teaching laboratory strategies for both faculty and students. Indeed, the experiments offered here were often developed by students and all have been tested and verified in laboratory classes. We extend our heartfelt thanks to each and every student involved in this effort. In particular, the analytical chemistry experiments were designed by Dr. K. C. Swallow and the alkyne hydration and empirical formula experiments were designed by Dr. Javier Horta, both of Merrimack College, North Andover, Massachusetts. We are most grateful to be able to include them here.

We are very appreciative of Taylor & Francis, especially the chemistry acquisitions editor Hilary Rowe, for giving us the opportunity to compile this series of experiments. Unlike a microwave reaction, the book has taken a bit of time to reach completion. We thank Hilary and our project coordinator Jennifer Ahringer for accommodating a number of requests for "just a bit more time." We also give a special thank you to our families who have been supportive of us throughout the design and writing of the book.

Neither of us would be writing this book, nor would we be so deeply involved in microwave chemistry if it were not for the students who have worked in our laboratories. Their enthusiasm, good ideas, and willingness to "give it a try" when we suggested something are greatly appreciated. Most especially, we have been incredibly fortunate to have close relationships with the major microwave equipment manufacturers. Of particular mention is CEM Corporation that not only supported us by providing equipment but also got us started in the area of undergraduate experiment development, publishing a series of experiments we developed in 2006.

We hope you enjoy our work, learn from the contents, and become avid microwavers!

**Nicholas E. Leadbeater and Cynthia B. McGowan**

# About the Authors

**Nicholas E. Leadbeater, PhD,** is an associate professor of chemistry at the University of Connecticut. A native of the United Kingdom, he graduated from the University of Nottingham, completed his doctorate in inorganic chemistry at the University of Cambridge, and served as a research fellow there for three years before joining the faculty of King's College, London. He moved to his current position at the University of Connecticut in 2004. Dr. Leadbeater's research interests are focused around development of new synthetic methodology, with an emphasis on cleaner, greener routes to known and novel compounds. A focus of the group's recent research effort has been the use of microwave heating and continuous-flow processing as enabling technologies. Dr. Leadbeater has a passion for undergraduate education, both developing new lecture and laboratory classes and incorporating undergraduate students into research. He was the recipient of the 2010 University of Connecticut College of Liberal Arts and Sciences Excellence in Teaching Award in the Physical Sciences.

**Cynthia B. McGowan, PhD,** is a full professor of chemistry at Merrimack College, North Andover, Massachusetts. Dr. McGowan graduated from Russell Sage College, completed her doctorate in organic chemistry at Brandeis University, and worked for a number of years as an industrial chemist before joining the faculty of Wellesley College prior to her current position at Merrimack College. A committed and popular undergraduate teacher, recognized by her peers with a teaching excellence award in 1999, she continues to adapt her material to the ever-changing world of technology so that her students are well prepared for graduate work or positions in industry. Her pioneering work in the use of microwave technology for organic chemistry experiments and teaching is "student-tested" and refined. Dr. McGowan is surrounded by a family of chemists (husband, daughters) and believes that giving students a serious and meaningful science experience can be powerful in helping with future career choices.

# About the Authors

# Introduction: Using Microwave Heating in Chemistry: The Basics*

**LEARNING GOALS**

- To gain a knowledge of the origins of microwave heating
- To understand how and why to perform a reaction using microwave heating
- To appreciate the range of chemistry possible when using microwave heating

## INTRODUCTION

The microwave oven is a time-saver in the kitchen at home. It is possible to heat things quickly and easily. The observation that microwave energy can heat food was first made by accident. Percy Spencer, an employee of the Raytheon Corporation was working on the development of radar equipment when he discovered that the candy bar in his pocket had melted. The next day he brought some popcorn from home and placed it close to the radar equipment and found that it quickly popped. Development of the microwave oven grew out of these observations and, by 1947 Raytheon was selling them. The original microwave ovens were about six feet (1.8 m) high and weighed in excess of 750 lb (340 kg). They cost $2,000–$3,000 to buy. The first popular home microwave oven was launched in 1967 and it is now estimated that over 200 million microwave ovens are in use in homes around the world.

Just as microwaves can heat food quickly, they can also be used for heating chemical reactions. Chemists in academic and industrial laboratories are finding that reactions can be performed quickly, easily, and effectively when using microwave heating. Although the pioneers in the field used microwave ovens they had taken into the lab from home, most experiments today are performed using scientific microwave apparatus. It is similar in many respects to its domestic microwave oven cousin. The microwaves are generated in the same way and at the same frequency. The big difference is that scientific microwave apparatus is designed with chemistry in mind and has a range of safety measures in place. It is also

* Parts contributed by Jason R. Schmink.

possible to monitor a range of parameters while reactions are running, including temperature, pressure, and microwave power.

## PHYSICAL CHEMISTRY CONCEPTS

The microwave region of the electromagnetic spectrum is classified as that between 0.3 and 300 gigahertz (GHz). Because applications such as cell phones and air traffic control operate in this range, regulatory agencies allow equipment for industrial, scientific, and medical (ISM) use to operate at only five specific frequencies. Household microwave ovens and scientific microwave apparatus typically operate at 2.45 GHz (12.25-cm wavelength), with few exceptions. Compared to much of the electromagnetic spectrum, microwave irradiation is of relatively low energy (Figure I.1). As a result, microwave irradiation does not break chemical bonds; it simply makes molecules rotate. This is in stark contrast to ultraviolet light which, when it interacts with molecules, can break bonds giving rise to reactive intermediates and a whole branch of science called photochemistry.

Microwaves, like all electromagnetic radiation, comprise oscillating electric and magnetic fields. Microwave heating is based upon the ability of a particular substance to absorb microwave energy and convert the electromagnetic energy to heat (kinetic energy). The interaction of microwave energy with a molecule can be explained by analogy to baseball or cricket (Figure I.2). During the swing, the batter or batsman can be said to be "rotationally excited" and can deliver some amount of rotational force to the incoming pitch (delivery in cricket). At the point of impact the rotational energy is rapidly converted into translational energy of

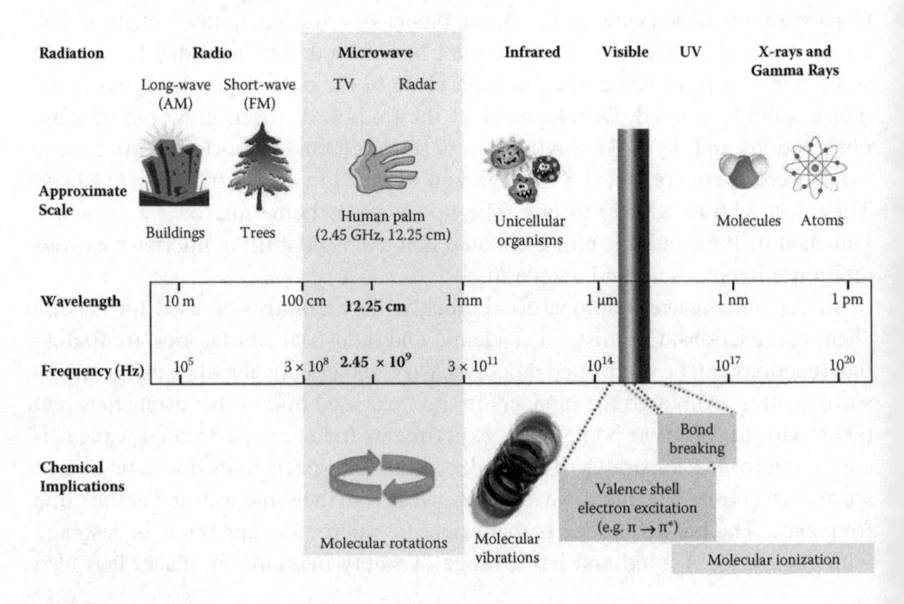


**FIGURE I.1** The electromagnetic spectrum.

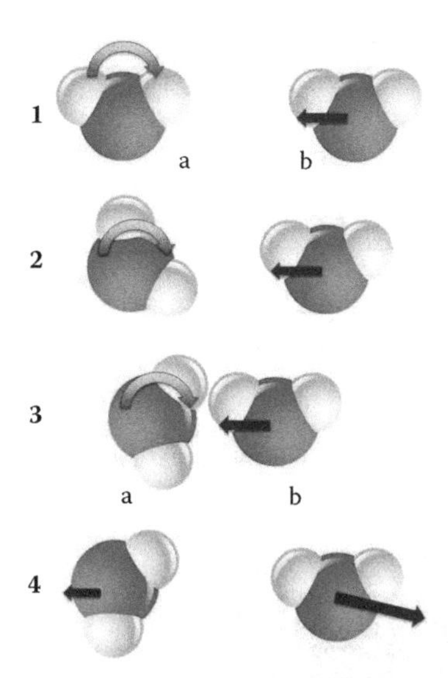


**FIGURE I.2** Molecule (a) that has been rotationally excited by microwave irradiation is approached by a second molecule (b), panels 1–3. Upon impact (panel 3), the rotational energy of (a) is converted to translational movement of (b). In panel 4, notice the increase in translational vector magnitude, the consequence of which leads to an increase in molecular collisions (kinetic energy). This concept is not unlike that of a baseball batter about to strike a baseball and impart the rotational energy onto a baseball in the form of translational energy.

the ball. Similarly, molecules with a dipole moment attempt to align themselves with the oscillating electric field of the microwave irradiation, leading to rotation. One water molecule excited rotationally by incident irradiation can strike a second molecule of water, converting rotational energy into translational energy. Under microwave irradiation, a large number of molecules are being rotationally excited and, as they strike other molecules, rotational energy is converted into translational energy (i.e., kinetic energy) and as a consequence heating is observed.

Each solvent or reagent will interact with microwave energy differently. Broadly speaking, the polarity of the solvent is a helpful tool for determining how well a reaction mixture will heat when irradiated with microwaves. Because microwave heating is dependent upon the dipole moment of a molecule, it stands to reason that more polar solvents such as dimethylsulfoxide, dimethylformamide, ethanol, and water better convert microwave irradiation into heat as compared to nonpolar solvents such as toluene or hexane.

## EQUIPMENT

There are two major types of microwave apparatus (Figures I.3 and I.4). The larger of these is termed a "multimode" unit; household microwave ovens are an example of this category. As the microwaves come into the cavity (heating

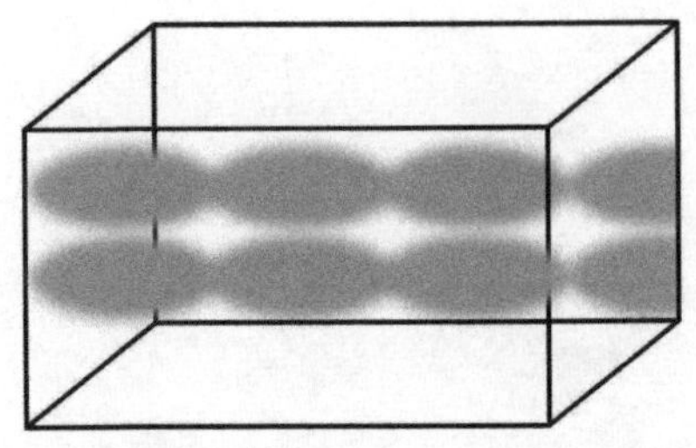

Areas of high and low microwave energy are found in the cavity of a multimode microwave unit

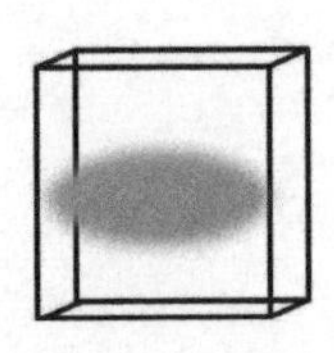

The cavity of a monomode microwave unit is designed to fit just one mode

**FIGURE I.3** Differences between multimode and monomode microwave units.

**FIGURE I.4** A (larger) multimode and a (smaller) monomode scientific microwave unit. (Reproduced with permission from CEM. Corporation.)

chamber), they will move around and bounce off the walls. As they do so, they will generate pockets (called modes) of high energy and low energy as the waves either reinforce or cancel each other out. As a result, the microwave field in the cavity is not uniform. There will instead be hot spots and cold spots, these corresponding to the pockets of high and low energy, respectively. This is why domestic microwave ovens (and their scientific cousins) often have turntables installed. The turntable allows for movement of food (or reaction vessels) around the cavity and equaling out the microwave energy to which the sample is exposed. It is possible to heat a number of reaction vessels at the same time by loading them onto a turntable and rotating them through the microwave field. When performing reactions on a small scale it is sometimes difficult to heat the mixture effectively in a multimode microwave unit. With the hot and cold spots that occur, it is hard to get constant microwave energy to heat a small sample. To overcome this, single-mode (also called monomode) units have been developed. The cavity of a monomode microwave system is the width of just one wave (mode) and it is possible to heat volumes as small as 0.2 mL effectively. The upper limit of monomode units is determined by the size of the microwave cavity and is usually in the region of 100 mL.

## MICROWAVE HEATING IN PREPARATIVE CHEMISTRY

Almost any reaction that requires heating can be performed using a microwave unit. There are a few exceptions, such as those reactions that are known to be highly exothermic. Most microwave reactions are performed in sealed glass tubes, capable of operating at temperatures up to 250°C and pressures of 300 psi (~20 bar) and it is possible to heat solvents to temperatures well in excess of their boiling points. Using microwave heating, reactions can often be performed in a matter of minutes. This is not because of anything special about microwave heating as compared to conventional heating but just that with a scientific microwave unit it is possible to access high temperatures in sealed vessels easily, quickly, and safely. Thus, a reaction that may take many hours to reach completion when performed at reflux using a hotplate or a steam, oil, or sand bath may be performed in five or ten minutes when heated in a sealed tube to a higher temperature using a microwave.

An alternative approach to running a reaction in a sealed tube is to perform it using an open round-bottom flask. This is useful when a reaction liberates a gas or if a product or by-product needs to be distilled out of the reaction as it is formed. However, when using an open-vessel approach one of the greatest attributes of microwave heating is eliminated, namely the ability to heat reactions to well above the normal boiling points of solvents in a safe and effective manner.

Microwave heating has found use in an industrial setting, especially at the research and development stage. Here, time is precious, so being able to run reactions quickly is a big advantage. In an interesting test performed at a pharmaceutical company, two scientists were told to make a series of compounds using the same suggested preparative route.* One of them used microwave heating and the other used conventional methods. After 37 days, the chemist using the conventional approach concluded that the molecules could not be generated using the route provided. The chemist using microwave heating optimized the reaction conditions and produced the final products in two days.

In the realm of organic chemistry, many classes of reaction are amenable to microwave heating, including: oxidations, reductions, substitutions, additions, cycloadditions (such as Diels–Alder reactions), rearrangements, ring-forming reactions, and metal-catalyzed reactions. Organic chemistry has perhaps been the area of most attention, however, it is possible to perform a wide range of other reactions using microwave heating. It has found application in materials and polymer chemistry and the preparation of inorganic compounds, as well as in the biosciences. It can also be used for extraction of metals or oils from samples, as well as for breaking down organic or inorganic compounds into smaller constituent parts, a technique known as digestion. An extension of this is application in the field of proteomics (the study of proteins and in particular their structure and function). Large complex proteins can be broken into smaller parts by

* Timesavings associated with microwave-assisted synthesis: A quantitative approach, C. R. Sarko in *Microwave Assisted Organic Synthesis* (J. P. Tierney and P. Lidstrom, Eds.), Blackwell Publishing, Oxford, 2005.

using microwave heating in conjunction with digestive enzymes such as trypsin. By knowing the constitution of these smaller parts, it is possible to piece together the sequence of the original protein.

## USE OF WATER AS A SOLVENT

Although there are certainly a number of reactions that do not tolerate the presence of water, for example, when using alkyl lithium reagents, there are many that not only tolerate water, but can benefit from its addition to the solvent system, or when it serves as the only solvent. Water is especially suitable for high-temperature organic reactions, and is frequently used in conjunction with microwave heating. The physical properties of water change as a function of temperature. It is characterized as a very polar solvent at room temperature, however, at elevated temperatures it becomes quite different. For example, water at 150°C has properties similar to dimethyl sulfoxide (DMSO). At 175°C water becomes similar to dimethyl formamide (DMF). At 200°C it is similar to acetonitrile, and water heated to 300°C has properties akin to acetone.* Water is thus able to dissolve, at least to a degree, organic reagents at high temperatures. Then, upon cooling, the products become insoluble and drop out of solution thereby making isolation of the newly synthesized compounds easy.

The use of water does potentially open avenues for performing "greener" chemistry. Green chemistry, also known as sustainable chemistry, is the design of chemical products and processes that reduce or eliminate the use or generation of hazardous substances. Water is inexpensive, nontoxic, nonflammable, and allows for easy product isolation. However, the true "greenness" of water needs to be considered on a case-by-case basis. Water cannot be incinerated after use as with organic solvents, and it takes a considerable amount of energy to distill water in order to purify it. Water purification at treatment plants is a costly and energy-intensive endeavor as well. This said, water still represents an attractive solvent and likely a greener choice than most if appropriate disposal treatments are employed. In addition, the ready access to elevated temperatures and the relatively efficient manner with which microwave irradiation heats water, makes microwave-assisted organic synthesis in water an attractive technique in the development of organic chemistry.

## PERFORMING A REACTION USING MICROWAVE HEATING

Monomode microwave units can process only one reaction vessel at a time. Using a sealed vessel, the reagents are loaded into a glass tube, a cap placed on the tube, and the tube placed inside the microwave cavity. A pressure sensor then connects with the tube, either automatically when the heating process is initiated or else manually before starting the reaction. Temperature is most often measured by means of an infrared sensor that is located in the base or the side

* Values for organic solvents obtained from: E. V. Anslyn and D. A. Dougherty, *Modern Physical Organic Chemistry*, University Science Books: Sausalito CA, 2006.

**FIGURE I.5** Examples of rotors used in a multimode microwave unit.

of the microwave cavity. This reads the temperature of the walls of the glass vessel without the thermometer having to be in contact with it. More accurate temperature measurement is possible by using a fiber-optic probe that is inserted directly into the reaction mixture, but this is more elaborate and often not used in a teaching laboratory setting.

Multimode microwave units can process multiple reaction vessels at a time. The vessels are placed in a rotor that sits inside the microwave cavity and rotates. Depending on the size of the reaction vessel used, anywhere from 4 to 192 individual sealed vessels can be loaded onto the rotor. Temperature is generally monitored and controlled by means of a fiber-optic probe inserted into one tube. This is termed the "control vessel". The temperature of other vessels may in some cases be monitored by use of an infrared sensor located in the wall or floor of the microwave cavity. Pressure is generally controlled by means of a special disk built into the tube cap. At pressures below the set limit of the disk, the vessel remains sealed. When the pressure approaches the limit of the vessel, the disk lifts slightly, venting the vessel. This can be thought of as similar to the operation of a pressure cooker in the kitchen.

Although a rotor approach (Figure I.5) could be applied to the development of chemistry by loading each vessel with a different permutation or combination of catalysts, reagents, and solvents, there are disadvantages to doing this. Inasmuch as temperature is controlled by the "control vessel," it could well be that all the other vessels are at different temperatures due to the difference in microwave absorptivity of the reaction mixtures. Therefore, a rotor approach is most commonly used to run multiple vessels of the same or similar reaction mixtures.

With both monomode and multimode microwave units, it is possible to stir reactions while they run. This is achieved by means of a magnetic stir bar placed in the reaction mixture, which couples with a magnetic stirrer motor located in the base of the microwave unit. It is always wise to stir a reaction while heating it in a microwave. This ensures good mixture of the reagents in the tube. Also, in cases where metal salts or other highly polar solids are used, inadvertent superheating of these very microwave-absorbing materials can be avoided.

When running a reaction there are two stages that are generally programmed into the unit:

(1) *Ramp time*: This is the time that the user wants the microwave to take to heat the reaction mixture up to the target temperature. Generally, a microwave power should be selected that affords a ramp time of 1–5°C/s to the target temperature. Use of too high an initial power in the ramp stage leads to inaccuracies, as the acquisition hardware and software are unable to keep pace. This situation often leads to temperature overshoot, sometimes by 10–20°C or more. It is generally best to provide too little power initially and adjust the applied power, if necessary.

(2) *Hold time*: This is the time that the reaction mixture is held at the target temperature before cooling back to room temperature. During this stage, the microwave power input will automatically fluctuate to hold the reaction mixture at the set temperature.

Upon completion of a heating program, the microwave unit will start a cooling process, generally blowing cool air over the reaction vessel or rotor. In the case of monomode microwave units comprising an automatically shutting cavity, the pressure sensor is usually programmed to release and the cavity open when the tube is cool enough to handle (around 50°C). With multimode microwave units, the apparatus is often set to cool for a period of time but at the end of this the reaction vessels may still be quite hot, depending on the temperature at which the reaction was performed and also the airflow around the rotor. (See Figure I.6.)

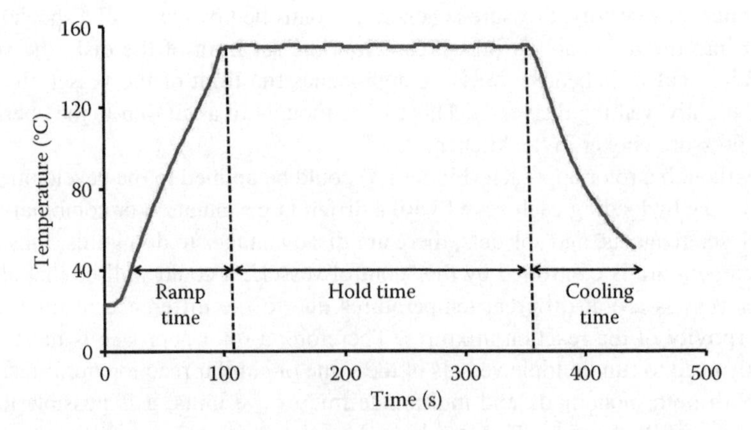


**FIGURE I.6** Example of a heating profile for a reaction performed using microwave heating.

## SAFETY

Scientific microwave apparatus is designed for preparative chemistry and is built with safety in mind. Domestic (household) microwave ovens should not be used for preparative chemistry. When employing microwave heating, all the safety precautions that are appropriate for performing a reaction using conventional heating should be adhered to, particularly the appreciation that reaction vessels can be hot and, when sealed, residual pressure needs to be released carefully at the end of the reaction. There are also some safety precautions that are specific to reactions using microwave heating:

- Adherence to the microwave manufacturer's user manual and guidelines is essential.
- Before running reactions in sealed vessels, it is important to check them for cracks or any other signs of damage.
- Only fill the reaction vessels to manufacturer's specifications; do not overfill the reaction vessels. An approximate rule of thumb is to fill the vessel to no farther than half its capacity.
- Only seal a closed reaction vessel with the manufacturer's recommended cap. These caps are designed to vent and re-seal in the case of an overpressurization during a reaction.
- If the cap is a twist-on type, be sure to use the appropriate tool to tighten the cap to the manufacturer-specified torque.
- It is important to monitor temperature and pressure profiles during the course of a reaction and to set safety limits before starting.
- Before performing a reaction at elevated temperatures, chemists should carefully consider the stability of the reagents and solvents they use at these temperatures.
- It is important to stir reaction mixtures throughout the heating process. The magnetic stir bars used should not be of exactly 3 cm in length because this equates to ¼ wavelength of a microwave at 2.45 GHz and thus acts as an antenna (leading to possible arcing and overheating).
- Upon completion of a heating cycle, it is important to check the temperature of the reaction vessel before removing it from the microwave cavity.
- When opening sealed vessels at the end of a reaction, be sure to point the vessel away from the face and any other people, preferably doing so in a hood with proper eye protection. Any remaining pressure will release as soon as the cap is removed. If the tube is very warm, it is advisable to cool it in an ice bath before removing the cap.

## EXPERIMENTAL PROTOCOLS

Although developed using the CEM Discover (monomode) and CEM MARS (multimode) microwave units, the experiments in this book can be performed on a range of commercially available scientific microwave units. With a few exceptions, each experiment has two procedures in it. The first is for use with monomode microwave apparatus and the second for use with multimode microwave units.

# 1 Diels–Alder Cycloaddition Reaction

**LEARNING GOALS**

- To prepare a six-membered carbocyclic ring by a Diels–Alder [2 + 4] cycloaddition reaction
- To determine the stereochemical outcome of a cycloaddition

## INTRODUCTION

A cycloaddition reaction involves two or more unsaturated molecules (or parts of the same molecule) combining with the formation of a cyclic adduct in which there is a net reduction of the bond multiplicity. When the unsaturated molecules consist of a conjugated diene and a species containing a carbon–carbon double or triple bond (termed a dienophile) the cycloaddition is referred to as a Diels–Alder reaction. The reaction is synthetically important because two new sigma bonds, a six-membered ring, and up to four stereocenters are formed in a single step.

trans cis R

conjugated diene dienophile

C* are new possible stereocenters

The yield of the reaction is usually greatest when the π-electron density of the diene and the dienophile are significantly different. This can be accomplished by using electron-rich dienes that are substituted with electron-donating groups (EDGs) and electron-poor dienophiles substituted with electron-withdrawing groups (EWGs). EDGs are often alkyl (–R), alkoxy (–OR), or siloxy ($-OSiR_3$) functionalities whereas the EWGs are often amide ($-CONR_2$), ester ($-CO_2R$), carboxylic acid ($-CO_2H$), aldehyde (–CHO), ketone (–COR), nitro ($-NO_2$), or nitrile (–CN) groups.

The reaction proceeds by a concerted mechanism where a single transition state controls the regiochemistry and stereochemistry of the product. The regiochemistry depends on the substitution pattern of the diene and the alkene. For the reaction of a monosubstituted diene and a monosubstituted alkene, if the EDG is on the carbon-1 position of the diene then the 1,2-disubstituted product is formed. When the EDG is on the carbon-2 position of the diene then the 1,4-disubstituted product is formed with none of the 1,3 analog.

EDG + EWG → EDG, EWG

*1,2-disubstituted product*

EDG + EWG → EDG, EWG [NOT EDG, EWG *1,3-disubstituted product*]

*1,4-disubstituted product*

The Diels–Alder reaction is also stereospecific relative to the diene and the dienophile. For example, if the dienophile is disubstituted the *cis*- or *trans*- relationship of these substituents is maintained in the product. Thus, in the examples below, the maleimide has a *cis*- stereochemistry of the hydrogen atoms and they are also *cis*- on the newly formed six-membered ring. Similarly, when *trans*-dimethyl fumarate is the dienophile, the product maintains the *trans*- relationship of the ester groups.

*malemide*

*dimethyl fumarate*

## PROCEDURE FOR USE IN A MONOMODE MICROWAVE UNIT

### Synthesis of 3a,4,9,9a-Tetrahydro-4-(Hydroxymethyl)-4,9[1′,2′]-Benzeno-1H-Benz[f]Isoindole-1,3(2H)-Dione

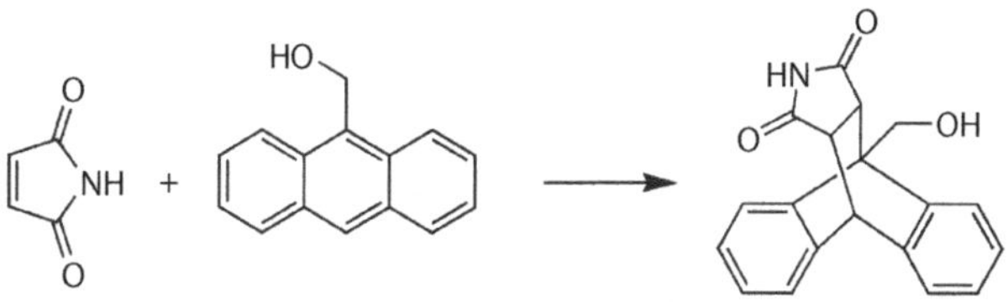

**Caution:** Maleimide and 9-anthracene methanol are toxic and irritants. The use of goggles with side-shields, lab coats, and gloves is considered minimum and nondiscretionary safety practice in the laboratory.

---

**Table of Reagents and Physical Constants**

| Reagent | Equiv. | FW | mmol | Mass (mg) | Density (g/mL) | Vol. (mL) | mp/bp°C |
|---|---|---|---|---|---|---|---|
| Maleimide (2,5-pyrroledione) $C_4H_3NO_2$ | 3 | 97.0 | 0.9 | 87 | — | — | 91–93 |
| 9-Anthracene methanol $C_{15}H_{12}O$ | 1 | 208 | 0.3 | 62 | — | — | 160–164 |
| Ethanol $C_2H_6O$ | — | — | — | — | — | 2.0 | 78 |
| Water $H_2O$ | — | — | — | — | — | 2.0 | 100 |

---

### Perform the Reaction

- ❑ In a 10-mL glass microwave reaction vessel place a magnetic stir bar, the maleimide, and the 9-anthracene methanol.
- ❑ Using a graduated cylinder, add the ethanol to the reaction vessel.
- ❑ Clamp the vessel over a stirring hotplate and stir the contents to dissolve the solids.
- ❑ Using a graduated cylinder, add the water to the reaction vessel (some precipitation may be observed).
- ❑ Seal the reaction vessel with a cap according to the microwave manufacturer's recommendations.
- ❑ Place the sealed reaction vessel in the microwave cavity.
- ❑ Program the microwave unit to heat the vessel contents to 150°C over a 2-minute ramp period and then hold at this temperature for 5 minutes.
- ❑ After the heating step is completed, allow the contents of the reaction vessel to cool to 50°C or below before removing it from the microwave cavity.

### Isolate the Product

- ❑ Carefully open the reaction vessel.
- ❑ Cool the reaction vessel in an ice bath for 10 minutes to complete the precipitation.
- ❑ While the solution is cooling set up a vacuum filtration system with a Hirsch funnel, side-arm flask, rubber collar, and a length of rubber vacuum tubing.
- ❑ In an ice bath, cool some water (8–10 mL) in a 25-mL Erlenmeyer flask for washing the precipitate after filtration.
- ❑ If a precipitate does not form, add a few drops of water until the solution looks cloudy and continue to cool in the ice bath.
- ❑ Connect the filtration system to a vacuum and place the correct size filter paper in the funnel.
- ❑ Wet the filter paper with a few drops of the cold water and start the vacuum to seal the filter paper in place.
- ❑ Filter the reaction mixture by pouring the contents into the funnel; transfer as much solid as possible.
- ❑ Wash the filter cake with two portions of ice-cold distilled water (3 mL).
- ❑ Allow the product to air dry on the filter funnel for several minutes.
- ❑ Transfer the filter cake to a large piece of filter paper to dry completely.

### Characterize the Product

- ❑ Weigh the dry product to determine the yield and calculate the yield and percent yield.
- ❑ Determine the melting point of the product and compare it to the literature value.
- ❑ To determine the purity of the product, run a thin-layer chromatography (TLC) of a sample using 50% ethyl acetate in hexane as eluent.
- ❑ If impure, re-crystallize the product using ethyl acetate as the solvent.
- ❑ Obtain an infrared (IR) spectrum of the product and compare it to that of the starting material.
- ❑ Obtain $^{1}$H-NMR and $^{13}$C-NMR spectra in $d_6$-acetone if instructed to do so.

## PROCEDURE FOR USE IN A MULTIMODE MICROWAVE UNIT

### SYNTHESIS OF 3A,4,9,9A-TETRAHYDRO-4-(HYDROXYMETHYL)-4,9[1′,2′]-BENZENO-1H-BENZ[f]ISOINDOLE-1,3(2H)-DIONE

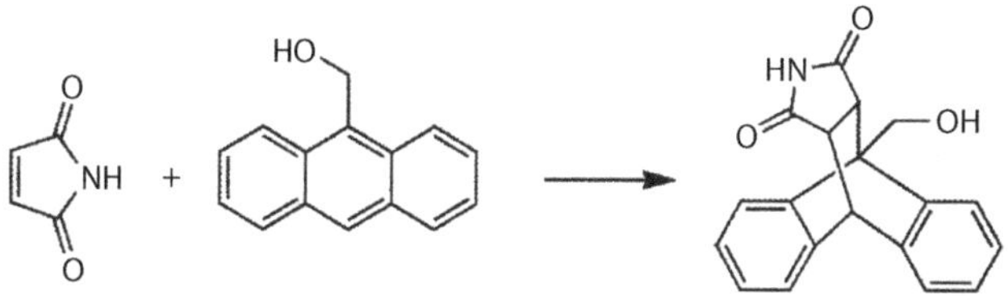

**Caution:** Maleimide and 9-anthracene methanol are toxic and irritants. The use of goggles with side-shields, lab coats, and gloves is considered minimum and nondiscretionary safety practice in the laboratory.

### Table of Reagents and Physical Constants

| Reagent | Equiv. | FW | mmol | Mass (mg) | Density (g/mL) | Vol. (mL) | mp/bp°C |
|---|---|---|---|---|---|---|---|
| Maleimide (2,5-pyrroledione) $C_4H_3NO_2$ | 3 | 97.0 | 0.9 | 87 | — | — | 91–93 |
| 9-Anthracene methanol $C_{15}H_{12}O$ | 1 | 208 | 0.3 | 62 | — | — | 160–164 |
| Ethanol $C_2H_6O$ | — | — | — | — | — | 2.0 | 78 |
| Water $H_2O$ | — | — | — | — | — | 2.0 | 100 |

### PERFORM THE REACTION

- ❑ In a 25-mL glass microwave reaction vessel place a magnetic stir bar, the maleimide, and the 9-anthracene methanol.
- ❑ Using a graduated cylinder, add the ethanol to the reaction vessel.
- ❑ Clamp the vessel over a stirring hotplate and stir the contents to dissolve the solids.
- ❑ Using a graduated cylinder, add the water to the reaction vessel (some precipitation may be observed).
- ❑ Seal the reaction vessel with a cap according to the microwave manufacturer's recommendations.
- ❑ Place the sealed reaction vessel in the carousel, noting the vessel's position number and ensuring that vessels are evenly spaced around the carousel.
- ❑ When all the group's reaction vessels are in place, load the carousel into the microwave cavity.

- ❑ If provided by the manufacturer, connect the temperature probe to the control vessel.
- ❑ Program the microwave unit to heat the vessel contents to 150°C over a 2-minute ramp period and then hold at this temperature for 5 minutes.
- ❑ After the heating step is completed, allow the contents of the reaction vessel to cool to 50°C or below before removing it from the microwave cavity.

### Isolate the Product

- ❑ Carefully open the reaction vessel.
- ❑ Cool the reaction vessel in an ice bath for 10 minutes to complete the precipitation.
- ❑ While the solution is cooling set up a vacuum filtration system with a Hirsch funnel, side-arm flask, rubber collar, and a length of rubber vacuum tubing.
- ❑ In an ice bath, cool some water (8–10 mL) in a 25-mL Erlenmeyer flask for washing the precipitate after filtration.
- ❑ If a precipitate does not form, add a few drops of water until the solution looks cloudy and continue to cool in the ice bath.
- ❑ Connect the filtration system to a vacuum and place the correct size filter paper in the funnel.
- ❑ Wet the filter paper with a few drops of the cold water and start the vacuum to seal the filter paper in place.
- ❑ Filter the reaction mixture by pouring the contents into the funnel; transfer as much solid as possible.
- ❑ Wash the filter cake with two portions of ice-cold distilled water (3 mL).
- ❑ Allow the product to air dry on the filter funnel for several minutes.
- ❑ Transfer the filter cake to a large piece of filter paper to completely dry.

### Characterize the Product

- ❑ Weigh the dry product to determine the yield and calculate the yield and percent yield.
- ❑ Determine the melting point of the product and compare it to the literature value.
- ❑ To determine the purity of the product, run a TLC of a sample using 50% ethyl acetate in hexane as eluent.
- ❑ If impure, re-crystallize the product using ethyl acetate as the solvent.
- ❑ Obtain an IR spectrum of the product and compare it to that of the starting material.
- ❑ Obtain $^{1}$H-NMR and $^{13}$C-NMR spectra in $d_6$-acetone if instructed to do so.

## QUESTIONS

1. Identify on the product the new stereocenters that are formed during the reaction.
2. How many stereoisomers would you predict for this Diels–Alder cycloaddition?
3. Predict the product of the following Diels–Alder reactions.

(a) 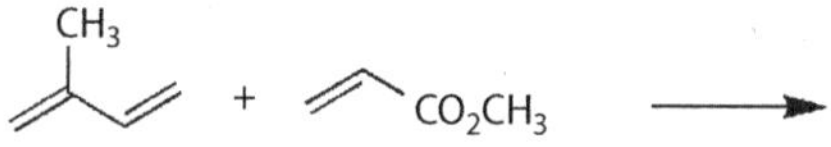 $CH_3$ + $CO_2CH_3$ ⟶

(b) 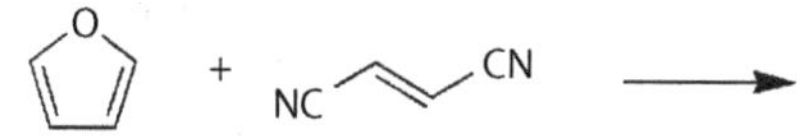 O + NC CN ⟶

(c) 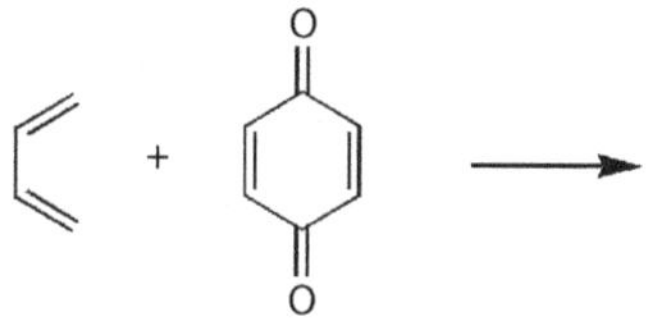 + O O ⟶

4. Predict the diene and dienophile that would lead to each of the following products.

(a) 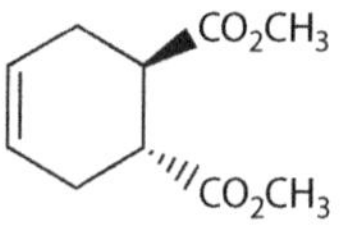 $CO_2CH_3$ $CO_2CH_3$

(b) 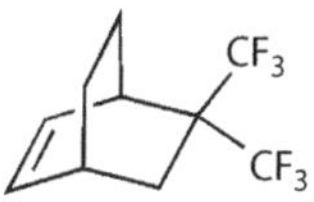 $CF_3$ $CF_3$

(c) 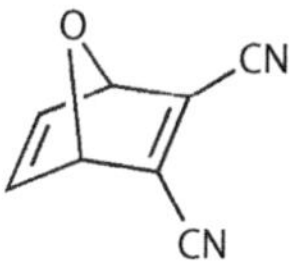 O CN CN

(d) 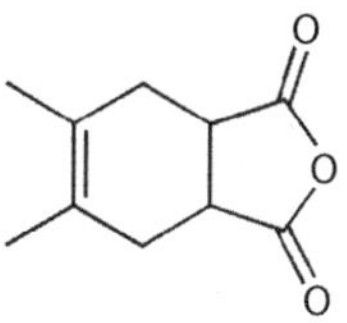 O O O

# 2 Second-Order Elimination Reaction

## *Preparation of Heptene from 2-Bromoheptane**

**LEARNING GOALS**

- To perform a base-induced dehydrohalogenation
- To investigate the effects of base size on the regiochemical outcome of the reaction
- To analyze the product mixture by gas chromatography

## INTRODUCTION

Eliminations are a class of reactions in which a multiple bond is formed in a substrate by means of loss of a small molecule. In the case of secondary or tertiary alkyl halides, alkenes are formed upon treatment with strong bases. This process is referred to as a dehydrohalogenation because the starting alkyl halide loses the halogen from one carbon (α-carbon) and a hydrogen atom from an adjacent carbon (β-carbon) to form a carbon–carbon double bond. This is termed a concerted reaction in that the base abstracts the hydrogen, the carbon–carbon double bond forms, and the halogen is eliminated simultaneously.

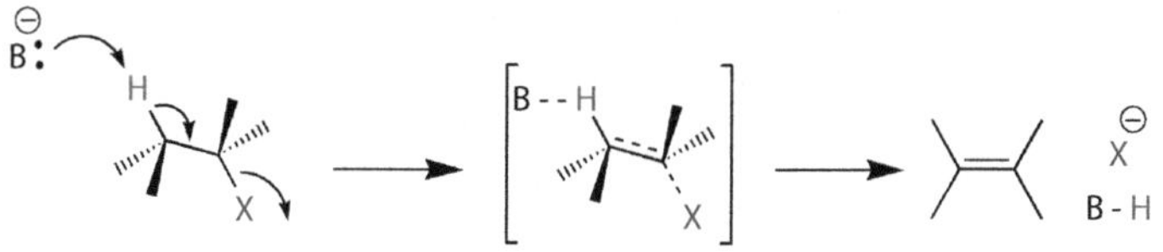

* Experimental procedure developed by Michael Luciano and David Foreman, Merrimack College. Modified from a conventional procedure: J. Mohrig, C. N. Hammond, P. F. Schatz, and T. C. Morrill, *Modern Projects and Experiments in Organic Chemistry: Miniscale and Standard Taper Microscale* (2nd ed.), Macmillan, London, 2002, pp. 77–81.

Nonsymmetrical secondary or tertiary alkyl halides, such as 2-bromoheptane, have more than one type of β-hydrogen and isomeric alkenes are isolated. The ratio of isomers produced depends on the size of the base (steric hindrance) and stability of the alkene products.

Saytzeff's (Zaitsev's) rule states that in an elimination reaction of an alkyl halide which has several different types of β-hydrogen, the more stable, highly substituted alkene isomer usually predominates. The following order of alkene stability is observed.

$$R_2C=CR_2 > R_2C=CHR > \underset{trans}{RHC=CHR} > \underset{cis}{RHC=CHR} > RHC=CH_2 > H_2C=CH_2$$

Starting with 2-bromoheptane, it would be predicted that 2-heptene should be the major product over 1-heptene. Furthermore, the *trans* (*E*) isomer is more stable than the *cis* (*Z*) isomer. The 2-heptenes result from the base abstracting one of the internal β hydrogen from carbon-3 instead of the terminal hydrogen on carbon-1.

*abstraction leads to 1-heptene*

*abstraction leads to 2-heptene*

In this experiment two different bases are used, one small (sodium methoxide) and one large (potassium *tert*-butoxide). The isomeric ratio of products will be determined by gas chromatography. From the product ratio it can be determined if the size of the base affects the ease of abstraction of the internal β-hydrogen versus the less hindered, terminal β-hydrogen. Be sure to record the boiling points of the three possible alkene products prior to performing the experiment.

## PROCEDURE FOR USE IN A MONOMODE MICROWAVE UNIT

Br

MW

*base*

**Caution:** Gloves should be worn while measuring out the reagents. 2-Bromoheptane, methanol, *tert*-butanol, and pentane are all flammable organic compounds and many are irritants. Sodium methoxide solution is flammable, corrosive, and moisture sensitive. Potassium *tert*-butoxide is corrosive and moisture sensitive. Keep the containers tightly closed when not in use. The use of goggles with side-shields, lab coats, and gloves is considered minimum and nondiscretionary safety practice in the laboratory.

### Table of Reagents and Physical Constants

| Reagent | Equiv. | FW | mmol | Mass (mg) | Density (g/mL) | Vol. (mL) | mp/bp°C |
|---|---|---|---|---|---|---|---|
| 2-Bromoheptane<br>$C_7H_{15}Br$ | 1 | 179 | 1.4 | 251 | 1.142 | 0.22 | 64–66 |
| ***Bases*** | | | | | | | |
| Sodium methoxide<br>25% solution in methanol<br>$CH_3ONa$ | 5 | 54.0 | 7.0 | 389 | 0.945 | 1.65 | — |
| Potassium *tert*-butoxide<br>$C_4H_9OK$ | 5 | 112 | 7.0 | 785 | — | — | — |
| ***Solvents*** | | | | | | | |
| Methanol, anhydrous<br>$CH_4O$ | — | — | — | — | — | 0.35 | 64.7 |
| *tert*-Butanol, anhydrous<br>$C_4H_{10}O$ | — | — | — | — | — | 2.0 | 23–26/83 |

### PERFORM THE REACTION

### A. Sodium Methoxide

- ❑ Dry (in an oven or with a heat gun) a 10-mL glass microwave reaction vessel in which a magnetic stir bar has been placed.
- ❑ Cool the vessel and stir bar to room temperature in a dry box or desiccator.
- ❑ Using automatic delivery pipettes add the 25% sodium methoxide solution, anhydrous methanol, and 2-bromoheptane to the reaction vessel.

- ❑ Seal the reaction vessel with a cap according to the microwave manufacturer's recommendations.
- ❑ Place the sealed reaction vessel in the microwave cavity.
- ❑ Program the microwave unit to heat the vessel contents to 120°C using an initial microwave power of 250 W and hold at this temperature for 4 minutes.
- ❑ After the heating step is completed, allow the contents of the reaction vessel to cool to 50°C or below before removing it from the microwave cavity.

### B. Potassium *tert*-Butoxide

- ❑ Dry (in an oven or with a heat gun) a 10-mL glass microwave reaction vessel in which a magnetic stir bar has been placed.
- ❑ Cool the vessel and stir bar to room temperature in a dry box or desiccator.
- ❑ Add *tert*-butanol to the reaction vessel using a graduated cylinder.
- ❑ Add the potassium *tert*-butoxide to the reaction vessel.
- ❑ Using an automatic delivery pipette add the 2-bromoheptane.
- ❑ Seal the reaction vessel with a cap according to the microwave manufacturer's recommendations.
- ❑ Place the sealed reaction vessel into the microwave cavity.
- ❑ Program the microwave unit to heat the vessel contents to 120°C using an initial microwave power of 250 W and hold at this temperature for 4 minutes.
- ❑ After the heating step is completed, allow the contents of the reaction vessel to cool to 50°C or below before removing it from the microwave cavity.

## Isolate the Product (A or B)

- ❑ Carefully open the reaction vessel.
- ❑ Remove the stirring bar from the solution using a magnetic retrieving wand.
- ❑ Add to the reaction vessel pentane (1.5 mL) and water (1.0 mL) and recap the vessel.
- ❑ Agitate the solution to dissolve the white solid.
- ❑ Carefully open the vessel and allow the layers to separate.
- ❑ Remove the lower aqueous layer using a Pasteur pipette and transfer it to a 25-mL Erlenmeyer flask.
- ❑ Add distilled water (1.0 mL) to the organics in the reaction vessel.
- ❑ Cap the vessel and mix the two layers.
- ❑ Allow the layers to separate.
- ❑ Remove the lower aqueous layer using a Pasteur pipette and transfer the aqueous layer to the Erlenmeyer flask as before.
- ❑ Extract the organic layer with additional water (1 mL) as before.
- ❑ Remove the aqueous layer to the Erlenmeyer flask.
- ❑ Add anhydrous sodium sulfate (300 mg) to the reaction vessel to dry the organic layer.
- ❑ Cap the vessel and allow the solution to stand for 5 minutes.

- ❑ Using a filter pipette move the organic solution from the reaction vessel into a small test tube or vial.
- ❑ Cap the test tube or vial.

### Analyze the Product (A or B)

- ❑ Inject 1 mL of the reaction solution into the injector port of the gas chromatograph equipped with a 30 m × 0.53 m × 3.0 m 100% dimethylpolysiloxane glass capillary column, helium carrier gas, and a thermal conductivity detector. The program is isothermal at 40°C.
- ❑ Obtain the chromatogram.
- ❑ Determine the area of each peak and calculate the relative composition of the mixture.
- ❑ Assign the peaks to each alkene based on their boiling points.

## PROCEDURE FOR USE IN A MULTIMODE MICROWAVE UNIT

Br

MW

*base*

**Caution:** Gloves should be worn while measuring out the reagents. 2-Bromoheptane, methanol, *tert*-butanol, and pentane are all flammable organic compounds and many are irritants. Sodium methoxide solution is flammable, corrosive, and moisture sensitive. Potassium *tert*-butoxide is corrosive and moisture sensitive. Keep the containers tightly closed when not in use. The use of goggles with side-shields, lab coats, and gloves is considered minimum and nondiscretionary safety practice in the laboratory.

### Table of Reagents and Physical Constants

| Reagent | Equiv. | FW | mmol | Mass (mg) | Density (g/mL) | Vol. (mL) | mp/bp°C |
|---|---|---|---|---|---|---|---|
| 2-Bromoheptane $C_7H_{15}Br$ | 1 | 179 | 4.22 | 0.754 | 1.142 | 0.66 | 64–66 |
| ***Bases*** | | | | | | | |
| Sodium methoxide 25% solution in methanol $CH_3ONa$ | 4.25 | 54.0 | 18.02 | — | 0.945 | 4.95 | — |
| Potassium *tert*-butoxide $C_4H_9OK$ | 5 | 112 | 18 | 2.0 | — | — | — |
| ***Solvents*** | | | | | | | |
| Methanol, anhydrous $CH_4O$ | — | 32 | — | — | 0.791 | 1.05 | 65 |
| *tert*-Butanol, anhydrous $C_4H_{10}O$ | — | 74 | — | — | 0.775 | 6.0 | 23–26/83 |

#### Perform the Reaction

### A. Sodium Methoxide

- ❑ Dry (in an oven or with a heat gun) a 25-mL glass microwave reaction vessel in which a magnetic stir bar has been placed.
- ❑ Cool the vessel and stir bar to room temperature in a dry box or desiccator.
- ❑ Using automatic delivery pipettes add the 25% sodium methoxide solution, anhydrous methanol, and 2-bromoheptane to the reaction vessel.

- ❑ Seal the reaction vessel with a cap according to the microwave manufacturer's recommendations.
- ❑ Place the sealed reaction vessel in the carousel, noting the vessel's position number and ensuring that vessels are evenly spaced around the carousel.
- ❑ When all the group's reaction vessels are in place, load the carousel into the microwave cavity.
- ❑ If provided by the manufacturer, connect the temperature probe to the control vessel.
- ❑ Program the microwave unit to heat the vessel contents to 120°C using an initial microwave power of 600 W and hold at this temperature for 6 minutes.
- ❑ After the heating step is completed, allow the contents of the reaction vessel to cool to 50°C or below before removing it from the microwave cavity.

### B. Potassium *tert*-Butoxide

- ❑ Dry (in an oven or with a heat gun) a 25-mL glass microwave reaction vessel in which a magnetic stir bar has been placed.
- ❑ Cool the vessel and stir bar to room temperature in a dry box or desiccator.
- ❑ Add *tert*-butanol to the reaction vessel using a graduated cylinder.
- ❑ Add the potassium *tert*-butoxide to the reaction vessel.
- ❑ Using an automatic delivery pipette add the 2-bromoheptane.
- ❑ Seal the reaction vessel with a cap according to the microwave manufacturer's recommendations.
- ❑ Place the sealed reaction vessel in the carousel, noting the vessel's position number and ensuring that vessels are evenly spaced around the carousel.
- ❑ When all the group's reaction vessels are in place, load the carousel into the microwave cavity.
- ❑ If provided by the manufacturer, connect the temperature probe to the control vessel.
- ❑ Program the microwave unit to heat the vessel contents to 120°C using an initial microwave power of 600 W and hold at this temperature for 6 minutes.
- ❑ After the heating step is completed, allow the contents of the reaction vessel to cool to 50°C or below before removing it from the microwave cavity.

## Isolate the Product (A or B)

- ❑ While the reaction is cooling, obtain two 25-mL Erlenmeyer flasks and label them "aqueous phase" and "organic phase".
- ❑ Carefully open the cooled reaction vessel.
- ❑ Transfer the solution to a 30-mL or 60-mL separatory funnel clamped to a ring stand.
- ❑ Rinse the reaction vessel with pentane (3 mL).
- ❑ Transfer the pentane solution to the separatory funnel.
- ❑ Add distilled water (3 mL) to the separatory funnel.
- ❑ Carefully stopper the separatory funnel and invert the funnel.

- ❑ Immediately vent the funnel by opening the stopcock to reduce pressure that may have developed in the funnel.
- ❑ Close the stopcock and mix the two layers several times by inverting the funnel repeatedly.
- ❑ Vent the funnel.
- ❑ Close the stopcock and re-clamp the funnel to a ring stand and remove the stopper.
- ❑ Allow the layers to separate.
- ❑ Remove the bottom aqueous layer through the stopcock into the Erlenmeyer flask labeled "aqueous phase".
- ❑ Wash the organic layer by adding distilled water (3 mL) to the separatory funnel.
- ❑ Carefully stopper the separatory funnel and invert the funnel.
- ❑ Immediately vent the funnel by opening the stopcock to reduce pressure that may have developed in the funnel.
- ❑ Close the stopcock and mix the two layers several times by inverting the funnel repeatedly.
- ❑ Vent the funnel.
- ❑ Close the stopcock and re-clamp the funnel to a ring stand and remove the stopper.
- ❑ Allow the layers to separate.
- ❑ Remove the bottom aqueous layer through the stopcock into the Erlenmeyer flask labeled "aqueous phase".
- ❑ Repeat twice the wash of the organic layer in the separatory funnel with additional distilled water (3 mL), collecting the bottom layer in the Erlenmeyer flask labeled "aqueous phase" each time.
- ❑ Pour the organic phase into the Erlenmeyer flask labeled "organic phase".
- ❑ Dry the organics by adding a drying reagent (500 mg) to the Erlenmeyer flask [either anhydrous magnesium sulfate ($MgSO_4$) or anhydrous sodium sulfate ($Na_2SO_4$)].
- ❑ Allow the organic solution containing the drying agent to stand for 5 minutes.
- ❑ Using a filter pipette move the organic solution from the reaction vessel into a test tube or vial.
- ❑ Cap the test tube or vial.

## Analyze the Product (A or B)

- ❑ Inject 1 mL of the reaction solution into the injector port of the gas chromatograph equipped with a 30 m × 0.53 m × 3.0 m 100% dimethylpolysiloxane glass capillary column, helium carrier gas, and a thermal conductivity detector. The program is isothermal at 40°C.
- ❑ Obtain the chromatogram.
- ❑ Determine the area of each peak and calculate the relative composition of the mixture.
- ❑ Assign the peaks to each alkene based on their boiling points.

## QUESTIONS

1. How does the size of the base affect the product composition in the E2 reaction?
2. Explain why the *trans* isomer of 2-heptene is more thermodynamically stable than the *cis* isomer.
3. In an E2 elimination, the hydrogen and bromine must be arranged anti-coplanar. Draw a 3D projection for the elimination of the internal β-hydrogen atoms and the bromine atom that leads to the formation of the *trans*- and *cis*-2-heptenes.
4. Give the structure of all the alkenes that can be formed and identify the major product if each of the following alkenes is reacted with sodium methoxide in methanol:
   a. 2-Bromopentane
   b. 3-Bromopentane
   c. 2-Bromo-3-methylbutane
   d. 2-Bromo-2-methylbutane

# 3 An Addition–Elimination Sequence

## *Preparation of a Bromoalkene**

**LEARNING GOALS**

- To perform a two-step one-pot reaction
- To use extraction as a product isolation technique
- To determine purity by thin-layer chromatography
- To purify a product by microchromatography

## INTRODUCTION

Two classes of reaction that are studied extensively in organic chemistry classes are additions and eliminations. In some ways they can be considered as opposites. Addition reactions involve the addition of a small molecule to a substrate. They are typical of unsaturated organic compounds such as alkenes and alkynes and may be considered as a process by which the double or triple bonds are fully or partially broken in order to accommodate additional atoms or groups of atoms in the molecule. For example, halogens undergo addition reactions with alkenes to yield di-halide products. Elimination reactions involve the loss of a small molecule from a substrate. They are the primary route by which organic compounds containing only single carbon–carbon bonds are transformed to compounds containing double or triple carbon–carbon bonds. Using these two processes in combination results in an overall substitution reaction.

$X_2$ *addition* → −HY *elimination* →

*substitution*

* Modified from a conventional procedure: C. Kueng, et al., *Synthesis*, 2005, 1319–1325.

In this experiment the starting alkene is a substituted cinnamic acid. The overall reaction will replace the carboxylic acid functionality with a bromine atom. The reaction uses *N*-bromosuccinimide (NBS) as a source of bromine, lithium acetate as a base, and is performed in an aqueous acetonitrile as the solvent.

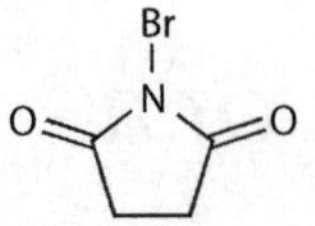

*N-bromosuccinamide*

Initially, an acid-base reaction takes place between the lithium acetate and the carboxylic acid. The π-electrons of the alkene then react with the bromine to form a bromonium ion. Under the aqueous conditions used, the bromonium ion will be in equilibrium with a benzylic carbocation. Finally, loss of carbon dioxide results in the formation of the product.

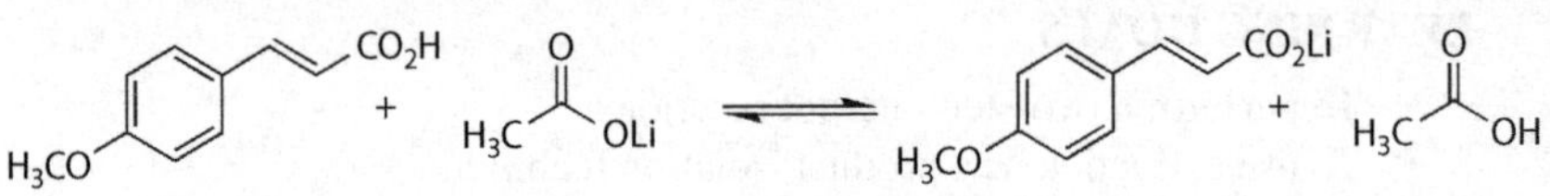

# PROCEDURE FOR USE IN A MONOMODE MICROWAVE UNIT

## Preparation of β-Bromo-*p*-Methoxystyrene (1-[(*E*)-2-Bromoethenyl]-4-Methoxybenzene)

$H_3CO$–$C_6H_4$–CH=CH–$CO_2H$ $\xrightarrow[\text{NBS/CH}_3\text{CO}_2\text{Li, H}_2\text{O/CH}_3\text{CN}]{\text{MW}}$ $H_3CO$–$C_6H_4$–CH=CH–Br

**Caution:** Acetonitrile and 4-methoxycinnamic acid are classified as irritants and skin contact should be avoided. *N*-bromosuccinimide (NBS) is corrosive. If the NBS appears yellow to orange, bromine has been liberated and is toxic and should be purified before using. The use of goggles with side-shields, lab coats, and gloves is considered minimum and nondiscretionary safety practice in the laboratory.

### Table of Reagents and Physical Constants

| Reagent | Equiv. | FW | mmol | Mass (mg) | Vol. (mL) | mp/bp°C |
|---|---|---|---|---|---|---|
| 4-Methoxycinammic acid $C_{10}H_{10}O_3$ | 2 | 178 | 0.65 | 116 | — | 173.5 |
| *N*-Bromosuccinimide (NBS) $C_4H_4BrO_2$ | 2.1 | 177 | 0.68 | 121 | — | 175–180 |
| Lithium acetate $LiC_2H_3O_2$ | 1 | 65.9 | 0.33 | 21 | — | 283–285 |
| 96% Aqueous acetonitrile $C_2H_3N$ | — | — | — | — | 1.0 | — |

## Perform the Reaction

- ❑ In a 25-mL Erlenmeyer flask, prepare a 96% aqueous acetonitrile solution by combining water (0.4 mL) and acetonitrile (9.6 mL).
- ❑ Stopper the flask.
- ❑ In a 10-mL glass microwave reaction vessel place a magnetic stir bar, the 4-methoxycinnamic acid, N-bromosuccinimide, and lithium acetate.
- ❑ Using a graduated cylinder, add the 96% aqueous acetonitrile solution.
- ❑ Seal the reaction vessel with a cap according to the microwave manufacturer's recommendations.
- ❑ Place the sealed reaction vessel in the microwave cavity.
- ❑ Program the microwave unit to heat the vessel contents to 100°C using an initial microwave power of 100 W and hold at this temperature for 4 minutes.
- ❑ After the heating step is completed, allow the contents of the reaction vessel to cool to 50°C or below before removing it from the microwave cavity.

## Isolate the Product

- ❑ While the reaction is cooling, obtain two 25-mL Erlenmeyer flasks and label them "aqueous phase" and "organic phase".
- ❑ Carefully open the reaction vessel.
- ❑ Add ethyl acetate (4 mL) to the reaction mixture.
- ❑ Clamp a 30-mL separatory funnel to a ring stand.
- ❑ Use a Pasteur pipette to transfer the reaction mixture to the separatory funnel leaving the magnetic stir bar in the reaction vessel.
- ❑ Rinse the reaction vessel with ethyl acetate (1 mL).
- ❑ Pipette the rinse into the separatory funnel that contains the reaction mixture.
- ❑ Add water (5 mL) to the separatory funnel.
- ❑ Carefully stopper the separatory funnel and invert the funnel.
- ❑ Immediately vent the funnel by opening the stopcock to release pressure that may have developed.
- ❑ Close the stopcock and mix the two layers several times by inverting the funnel repeatedly.
- ❑ Vent the funnel as before.
- ❑ Close the stopcock, re-clamp the separatory funnel to the ring stand, and remove the stopper.
- ❑ Allow the layers to separate.
- ❑ Move the bottom aqueous layer through the stopcock into the 25-mL Erlenmeyer flask labeled "aqueous phase".
- ❑ Add 6 M sodium hydroxide solution (5 mL) to the separatory funnel that contains the upper organic phase.
- ❑ Stopper the funnel and mix the two phases by inverting the funnel and releasing the pressure, as before.
- ❑ Re-clamp the funnel and remove the stopper.
- ❑ Move the bottom aqueous phase to the Erlenmeyer flask labeled "aqueous phase".
- ❑ Add distilled water (5 mL) to the separatory funnel containing the organic layer.
- ❑ Stopper the funnel and mix the two phases by inverting the funnel and releasing the pressure, as before.
- ❑ Re-clamp the funnel and remove the stopper.
- ❑ Move the bottom aqueous phase to the Erlenmeyer flask labeled "aqueous phase".
- ❑ Add saturated sodium chloride solution (5 mL) to the organic phase in the separatory funnel.
- ❑ Stopper the funnel and mix the two phases by repeatedly inverting the separatory funnel, venting to release any pressure as before.
- ❑ Re-clamp the separatory funnel and remove the stopper.
- ❑ Move the bottom aqueous phase through the stopcock into the 25-mL flask labeled "aqueous phase".
- ❑ Pour the top organic layer from the separatory funnel into the Erlenmeyer flask labeled "organic phase".

- ❑ Add anhydrous magnesium sulfate ($MgSO_4$) to the flask containing the organic layer and stopper the flask.
- ❑ Allow the flask to sit for 10 minutes to complete the drying process.
- ❑ Tare a 25-mL or 50-mL round bottom flask while the organic layer is drying.
- ❑ Transfer the organic layer to the round bottom flask with a filter pipette; avoid transferring any of the drying agent.
- ❑ Rinse the Erlenmeyer flask and drying agent with ethyl acetate (3 mL) and add this solution to the solution in the round bottom flask.
- ❑ To determine the purity of the product, run a TLC of this solution by removing a small sample from the round bottom flask with a micropipette or TLC spotter.
- ❑ Spot the solution on a prepared TLC plate.
- ❑ Prepare a TLC chamber by taking a beaker large enough to fit the TLC plate and adding a small layer of hexane.
- ❑ Cover the beaker with a watch glass.
- ❑ Carefully place the TLC plate in the beaker leaning the plate against the side of the beaker.
- ❑ Be sure the spot is above the eluting solvent.
- ❑ Re-cover the beaker.
- ❑ Once the solvent front is near the top of the plate carefully remove the plate from the chamber.
- ❑ Immediately mark the solvent front with a pencil and then allow the plate to dry.
- ❑ Visualize the TLC plate with a UV lamp and mark any spots observed.
- ❑ Remove the ethyl acetate from the organic solution under reduced pressure until constant weight is observed.
- ❑ Re-weigh the flask containing the product; calculate the yield and percent yield.

## Purification of the Product if the TLC Shows Unreacted Starting Material

- ❑ Plug a 5.75-mm disposable glass pipette with a small piece of glass wool.
- ❑ Add to the pipette, silica gel to a height of 5 cm.
- ❑ Dissolve the solid in the round bottom flask in a minimal amount of hexane.
- ❑ Pass the hexane solution through the silica gel in the pipette into a clean-tared round-bottom or pear-shaped flask.
- ❑ After all of the solution has been added to the pipette, wash the silica gel with hexane (5 mL) collecting the solution in the same flask.
- ❑ Remove the hexane under reduced pressure until a constant weight is observed.
- ❑ Re-weigh the flask containing the white solid.
- ❑ Calculate the yield and percent yield.

## Characterize the Product

- ❑ Obtain an IR spectrum of the product and compare it with that of the starting material.
- ❑ Obtain $^{1}H$-NMR and $^{13}C$-NMR spectra in $CDCl_3$ if instructed to do so.

# PROCEDURE FOR USE IN A MULTIMODE MICROWAVE UNIT

## Preparation of β-Bromo-*p*-Methoxystyrene (1-[(*E*)-2-Bromoethenyl]-4-Methoxybenzene)

$H_3CO$–C₆H₄–CH=CH–$CO_2H$ $\xrightarrow[\text{NBS/CH}_3\text{CO}_2\text{Li} \;\; \text{H}_2\text{O/CH}_3\text{CN}]{\text{MW}}$ $H_3CO$–C₆H₄–CH=CH–Br

**Caution:** Acetonitrile and 4-methoxycinnamic acid are classified as irritants and skin contact should be avoided. *N*-Bromosuccinimide (NBS) is corrosive. If the NBS appears yellow to orange, bromine has been liberated and is toxic and should be purified before using. The use of goggles with side-shields, lab coats, and gloves is considered minimum and nondiscretionary safety practice in the laboratory.

### Table of Reagents and Physical Constants

| Reagent | Equiv. | FW | mmol | Mass (mg) | Vol. (mL) | mp/bp°C |
|---|---|---|---|---|---|---|
| 4-Methoxycinammic acid $C_{10}H_{10}O_3$ | 2 | 178 | 2.0 | 366 | — | 173.5 |
| *N*-Bromosuccinimide (NBS) $C_4H_4BrO_2$ | 2.1 | 177 | 2.1 | 374 | — | 175–180 |
| Lithium acetate $LiC_2H_3O_2$ | 1 | 65.9 | 1.0 | 66 | — | 283–285 |
| 96% Aqueous acetonitrile $C_2H_3N$ | — | — | — | — | 4.0 | — |

## Perform the Reaction

- ❑ In a 25-mL Erlenmeyer flask, prepare a 96% aqueous acetonitrile solution by combining water (0.4 mL) and acetonitrile (9.6 mL).
- ❑ Stopper the flask.
- ❑ In a 25-mL glass microwave reaction vessel place a magnetic stir bar, the 4-methoxycinnamic acid, N-bromosuccinimide, and lithium acetate.
- ❑ Using a graduated cylinder add the 96% aqueous acetonitrile solution.
- ❑ Seal the reaction vessel with a cap according to the microwave manufacturer's recommendations.
- ❑ Place the sealed reaction vessel in the carousel, noting the vessel's position number and ensuring that vessels are evenly spaced around the carousel.
- ❑ When all the group's reaction vessels are in place, load the carousel into the microwave cavity.

- ❑ If provided by the manufacturer, connect the temperature probe to the control vessel.
- ❑ Program the microwave unit to heat the vessel contents to 100°C using an initial microwave power of 300 W and hold at this temperature for 4 minutes.
- ❑ After the heating step is complete, allow the contents of the reaction vessel to cool to 50°C or below before removing it from the microwave cavity.

## Isolate the Product

- ❑ While the reaction is cooling, obtain two 50-mL Erlenmeyer flasks and label them "aqueous phase" and "organic phase".
- ❑ Carefully open the reaction vessel.
- ❑ Add ethyl acetate (8 mL) to the reaction mixture.
- ❑ Clamp a 60-mL separatory funnel to a ring stand.
- ❑ Use a Pasteur pipette to transfer the reaction mixture to the separatory funnel leaving the magnetic stir bar in the reaction vessel.
- ❑ Rinse the reaction vessel with ethyl acetate (1 mL).
- ❑ Pipette the rinse into the separatory funnel that contains the reaction mixture.
- ❑ Add water (8 mL) to the separatory funnel.
- ❑ Carefully stopper the separatory funnel and invert the funnel.
- ❑ Immediately vent the funnel by opening the stopcock to release pressure that may have developed.
- ❑ Close the stopcock and mix the two layers several times by inverting the funnel repeatedly.
- ❑ Vent the funnel as before.
- ❑ Close the stopcock, re-clamp the separatory funnel to the ring stand and remove the stopper.
- ❑ Allow the layers to separate.
- ❑ Move the bottom aqueous layer through the stopcock into a 25-mL Erlenmeyer flask labeled "aqueous phase".
- ❑ Add 6 M sodium hydroxide solution (8 mL) to the separatory funnel that contains the upper organic phase.
- ❑ Stopper the funnel and mix the two phases by inverting the funnel and releasing the pressure, as before.
- ❑ Re-clamp the funnel and remove the stopper.
- ❑ Move the bottom aqueous phase to the Erlenmeyer flask labeled "aqueous phase".
- ❑ Add distilled water (8 mL) to the separatory funnel containing the organic layer.
- ❑ Stopper the funnel and mix the two phases by inverting the funnel and releasing the pressure, as before.
- ❑ Re-clamp the funnel and remove the stopper.
- ❑ Move the bottom aqueous phase to the Erlenmeyer flask labeled "aqueous phase".

- ❑ Add saturated sodium chloride solution (8 mL) to the organic phase in the separatory funnel.
- ❑ Stopper the funnel and mix the two phases by repeatedly inverting the separatory funnel, venting to release any pressure as before.
- ❑ Re-clamp the separatory funnel and remove the stopper.
- ❑ Move the bottom aqueous phase through the stopcock into the 50-mL flask labeled "aqueous phase".
- ❑ Pour the top organic layer from the separatory funnel into the Erlenmeyer flask labeled "organic phase".
- ❑ Add anhydrous magnesium sulfate ($MgSO_4$) to the flask containing the organic layer and stopper the flask.
- ❑ Allow the flask to sit for 10 minutes to complete the drying process.
- ❑ Tare a 25-mL or 50-mL round bottom flask while the organic layer is drying.
- ❑ Transfer the organic layer to the round bottom flask with a filter pipette; avoid transferring any of the drying agent.
- ❑ Rinse the Erlenmeyer flask and drying agent with ethyl acetate (3 mL) and add this solution to the solution in the round bottom flask.
- ❑ To determine the purity of the product, run a TLC of this solution by removing a small sample from the round bottom flask with a micropipette or TLC spotter.
- ❑ Spot the solution on a prepared TLC plate.
- ❑ Prepare a TLC chamber by taking a beaker large enough to fit the TLC plate and adding a small layer of hexane.
- ❑ Cover the beaker with a watch glass.
- ❑ Carefully place the TLC plate in the beaker leaning the plate against the side of the beaker.
- ❑ Be sure the spot is above the eluting solvent.
- ❑ Re-cover the beaker.
- ❑ Once the solvent front is near the top of the plate carefully remove the plate from the chamber.
- ❑ Immediately mark the solvent front with a pencil and then allow the plate to dry.
- ❑ Visualize the TLC plate with a UV lamp and mark any spots observed.
- ❑ Remove the ethyl acetate from your organic solution under reduced pressure until constant weight is observed.
- ❑ Re-weigh the flask containing the product; calculate the yield and percent yield.

## Purification of the Product if the TLC Shows Unreacted Starting Material

- ❑ Plug a 5.75-mm disposable glass pipette with a small piece of glass wool.
- ❑ Add to the pipette, silica gel to a height of 5 cm.
- ❑ Dissolve the solid in the round bottom flask in a minimal amount of hexane.

- ❑ Pass the hexane solution through the silica gel in the pipette into a clean-tared round-bottom or pear-shaped flask.
- ❑ After all of the solution has been added to the pipette, wash the silica gel with hexane (5 mL) collecting the solution in the same flask.
- ❑ Remove the hexane under reduced pressure until a constant weight is observed.
- ❑ Re-weigh the flask containing the white solid.
- ❑ Calculate the yield and percent yield.

## Characterize the Product

- ❑ Obtain an IR spectrum of the product and compare it with that of the starting material.
- ❑ Obtain $^{1}$H-NMR and $^{13}$C-NMR spectra in $CDCl_3$ if instructed to do so.

## QUESTIONS

1. In your reaction, explain why the *trans* isomer of the product is formed.
2. How can you tell from the NMR spectrum or the IR spectrum that the *trans* isomer of the product is formed?
3. What is the purpose of using a 6 M sodium hydroxide solution during the extraction process?
4. Write the mechanism for the bromination of an alkene using elemental bromine.

# 4 Fischer Esterification

## *Preparation of Ethyl-4-Aminobenzoate (Benzocaine)*

**LEARNING GOALS**

- To understand the acid-catalyzed esterification reaction
- To separate a reactant from product by controlling solubility by pH
- To separate a solid product from a solution by vacuum filtration

## INTRODUCTION

Functional groups containing carbon-to-oxygen double bonds (a carbonyl, C = O) are found in most medicines, materials, and natural products. Within the carbonyl family there is a series of functional groups called the carboxylic acids and their derivatives. Carboxylic acids are organic molecules that contain the $-CO_2H$ functional group. The derivatives are any functional group that, when hydrolyzed with water, generate the carboxylic acid. There are five such carboxylic acid derivatives:

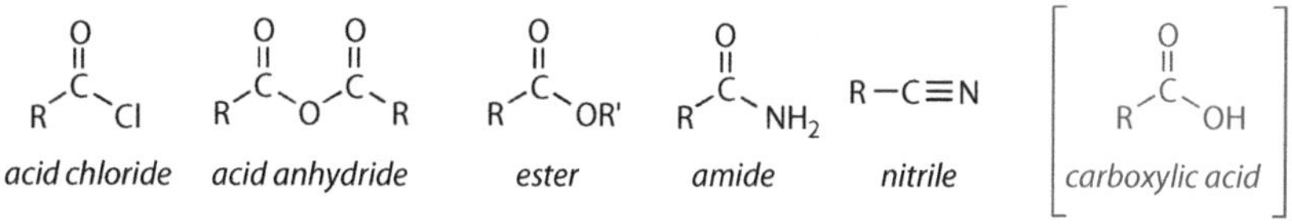

*acid chloride* *acid anhydride* *ester* *amide* R−C≡N *nitrile* *carboxylic acid*

Esters can be prepared by the condensation of a carboxylic acid with an alcohol. The low molecular-weight esters often have fruity odors and are used industrially for flavors and fragrances such as banana oil and wintergreen. The Fischer esterification reaction is the acid-catalyzed condensation of a carboxylic acid and alcohol.

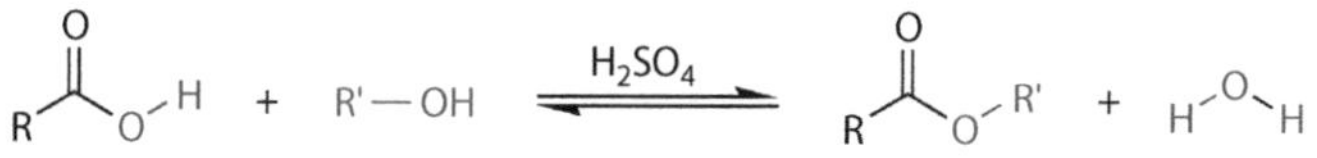

$$RCO_2H + R'{-}OH \underset{}{\overset{H_2SO_4}{\rightleftharpoons}} RCO_2R' + H_2O$$

The mechanism for the reaction is

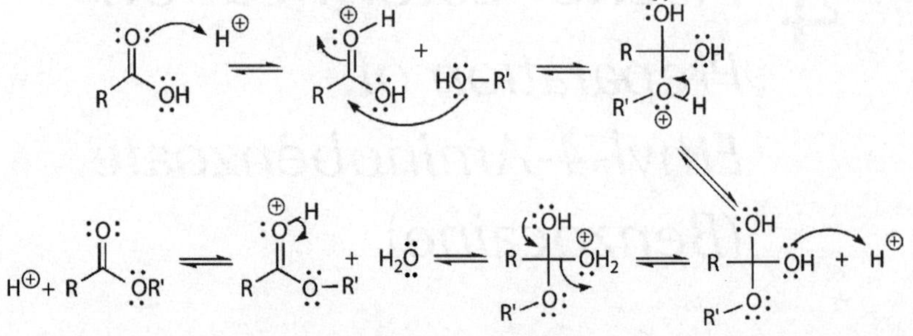

An understanding of Le Chatelier's principle is important when trying to improve the yield of this reversible reaction. To drive the reaction toward the ester product, one of the two reagents (either the alcohol or the carboxylic acid) can be used in a significant excess. This shifts the equilibrium to the left. In addition, a drying agent can be added to sequester the water as it is produced, essentially shutting down the reverse pathway.

## PROCEDURE FOR USE IN A MONOMODE MICROWAVE UNIT

$CO_2H$ / $NH_2$ (benzene ring) + $CH_3CH_2OH$ $\xrightarrow{H_2SO_4}$ $CO_2CH_2CH_3$ / $NH_2$ (benzene ring) + $H_2O$

**Caution:** The reaction requires the use of concentrated sulfuric acid, a strong acid that can cause serious burns if spilled on the skin. Transfer the sulfuric acid to a glass graduated cylinder and pipette the required amount directly into the reaction tube. Flush any extra reagent down the sink with large amounts of water. If any sulfuric acid is accidentally spilled on the skin, immediately wash the area with large amounts of water to prevent a burn. The use of goggles with side-shields, lab coats, and gloves is considered minimum and nondiscretionary safety practice in the laboratory.

### Table of Reagents and Physical Constants

| Reagent | Equiv. | FW | mmol | Mass (mg) | Density (g/mL) | Vol. (mL) | mp/bp°C |
|---|---|---|---|---|---|---|---|
| 4-Aminobenzoic acid $C_7H_7NO_2$ | 1 | 134 | 1.49 | 200 | 1.374 | — | 187–188 |
| Ethanol $C_2H_6O$ | 23 | 46.1 | 34 | 1580 | 0.789 | 2.0 | 78 |
| Sulfuric acid (conc.) $H_2SO_4$ | — | 98.1 | 3.7 | 366 | 1.83 | 0.2 | 324 |

### Perform the Reaction

- ❏ In a 10-mL glass microwave reaction vessel place a magnetic stir bar and the 4-aminobenzoic acid.
- ❏ Add the ethanol using either a graduated pipette or a 10-mL graduated cylinder.
- ❏ Stir the contents with the aid of a stirring hotplate to dissolve the solid.
- ❏ Carefully in a fume hood add the concentrated sulfuric acid dropwise.
- ❏ Seal the reaction vessel with a cap according to the microwave manufacturer's recommendations.
- ❏ Program the microwave unit to heat the vessel contents to 140°C over a period of 2 minutes and then hold at this temperature for 5 minutes.
- ❏ After the heating step is completed, allow the contents of the reaction vessel to cool to 50°C or below before removing it from the microwave cavity.

### Isolate the Product

- ❑ Place 10% aqueous potassium carbonate (10 mL) in a 50-mL Erlenmeyer flask.
- ❑ Cool the Erlenmeyer flask in an ice bath while the reaction vessel is in the microwave.
- ❑ Carefully open the reaction vessel.
- ❑ Add the contents of the reaction vessel dropwise, with swirling, to the carbonate solution in the Erlenmeyer flask.
- ❑ Check the pH of the solution by dipping the tip of a glass pipette into the solution and then touch the tip to broad-range pH paper.
- ❑ If the solution is still acidic add 10% potassium carbonate, 1 mL at a time, to the solution until the solution is neutral to basic.
- ❑ Continue cooling the solution in the ice bath for 10 minutes after the addition is complete. Solid material will precipitate out of the solution.*
- ❑ While the solution is cooling set up a vacuum filtration system with a Hirsch funnel, side-arm flask, rubber collar, and a length of rubber vacuum tubing.
- ❑ Cool distilled water (10 mL) in a 25-mL Erlenmeyer flask in an ice bath for washing the precipitate after filtration.
- ❑ Connect the filtration system to a vacuum and place the correct size filter paper in the funnel.
- ❑ Wet the filter paper with a few drops of the cold distilled water and start the vacuum to seal the filter paper in place.
- ❑ Filter the reaction mixture by pouring the contents of the 50-mL Erlenmeyer flask into the funnel; transfer as much solid as possible.
- ❑ Rinse the flask with cold distilled water (3 mL) and add it to the filter funnel.
- ❑ Rinse the precipitate with additional cold distilled water (3 mL).
- ❑ Allow the precipitate to dry on the funnel for 10 minutes.
- ❑ Transfer the solid to a large piece of filter paper to completely dry.
- ❑ Weigh the dry product and calculate the yield and percent yield.

### Characterize the Product

- ❑ To determine the purity of the product, run a TLC of a sample using ethyl acetate as eluent.
- ❑ Obtain the melting point of the product and compare it to the literature value.
- ❑ Obtain an IR spectrum of the product and compare it to that of the starting material.
- ❑ Obtain $^{1}$H-NMR and $^{13}$C-NMR spectra in $CDCl_3$ if instructed to do so.

---

* If a precipitate does not form, the product can be extracted from the aqueous solution with ethyl acetate.

## PROCEDURE FOR USE IN A MULTIMODE MICROWAVE UNIT

$$H_2N\text{-}C_6H_4\text{-}CO_2H + CH_3CH_2OH \xrightarrow{H_2SO_4} H_2N\text{-}C_6H_4\text{-}CO_2CH_2CH_3 + H_2O$$

**Caution:** The reaction requires the use of concentrated sulfuric acid, a strong acid that can cause serious burns if spilled on the skin. Transfer the sulfuric acid to a glass graduated cylinder and pipette the required amount directly into the reaction tube. Flush any extra reagent down the sink with large amounts of water. If any sulfuric acid is accidentally spilled on the skin, immediately wash the area with large amounts of water to prevent a burn. The use of goggles with side-shields, lab coats, and gloves is considered minimum and nondiscretionary safety practice in the laboratory.

### Table of Reagents and Physical Constants

| Reagent | Equiv. | FW | mmol | Mass (g) | Density (g/mL) | Vol. (mL) | mp/bp°C |
|---|---|---|---|---|---|---|---|
| 4-Aminobenzoic acid $C_7H_7NO_2$ | 1 | 134.1 | 2.2 | 0.300 | 1.374 | — | 187–188 |
| Ethanol $C_2H_6O$ | 46 | 46.1 | 102 | 4.7 | 0.789 | 6.0 | 78 |
| Sulfuric acid (conc.) $H_2SO_4$ | — | 98.1 | 7.5 | 0.732 | 1.83 | 0.4 | 324 |

### PERFORM THE REACTION

- ❑ In a 25-mL glass microwave reaction vessel place a magnetic stir bar and the 4-aminobenzoic acid.
- ❑ Add the ethanol using either a graduated pipette or a 10-mL graduated cylinder.
- ❑ Stir the contents with the aid of a stirring hotplate to dissolve the solid.
- ❑ Carefully in a fume hood add the concentrated sulfuric acid dropwise.
- ❑ Seal the reaction vessel with a cap according to the microwave manufacturer's recommendations.
- ❑ Place the sealed reaction vessel in the carousel, noting the vessel's position number and ensuring that vessels are evenly spaced around the carousel.
- ❑ When all the group's reaction vessels are in place, load the carousel into the microwave cavity.

- ❑ If provided by the manufacturer, connect the temperature probe to the control vessel.
- ❑ Program the microwave unit to heat the vessel contents to 140°C at 200 watts over a 2-minute ramp period and then hold at this temperature for 5 minutes.
- ❑ After the heating step is completed, allow the contents of the reaction vessel to cool to 50°C or below before removing it from the microwave cavity.

## Isolate the Product

- ❑ Place 10% aqueous potassium carbonate (20 mL) in a 50-mL Erlenmeyer flask.
- ❑ Cool the Erlenmeyer flask in an ice bath while the reaction vessel is in the microwave.
- ❑ Carefully open the reaction vessel.
- ❑ Add the contents of the reaction vessel dropwise, with swirling, to the carbonate solution in the Erlenmeyer flask.
- ❑ Check the pH of the solution by dipping the tip of a glass pipette into the solution and then touch the tip to broad-range pH paper.
- ❑ If the solution is still acidic add 10% potassium carbonate, 1 mL at a time, to the solution until the solution is neutral to basic.
- ❑ Continue cooling the solution in the ice bath for 10 minutes after the addition is complete. Solid material will precipitate out of the solution.*
- ❑ While the solution is cooling set up a vacuum filtration system with a Hirsch funnel, side-arm flask, rubber collar, and a length of rubber vacuum tubing.
- ❑ Cool distilled water (10 mL) in 25-mL Erlenmeyer flask in an ice bath for washing the precipitate after filtration.
- ❑ Connect the filtration system to a vacuum and place the correct size filter paper in the funnel.
- ❑ Wet the filter paper with a few drops of the cold distilled water and start the vacuum to seal the filter paper in place.
- ❑ Filter the reaction mixture by pouring the contents of the 50-mL Erlenmeyer flask into the funnel; transfer as much solid as possible.
- ❑ Rinse the flask with cold distilled water (3 mL) and add it to the filter funnel.
- ❑ Rinse the precipitate with additional cold distilled water (3 mL).
- ❑ Allow the precipitate to dry on the funnel for 10 minutes.
- ❑ Transfer the solid to a large piece of filter paper to completely dry.
- ❑ Weigh the dry product and calculate the yield and percent yield.

---

* If a precipitate does not form, the product can be extracted from the aqueous solution with ethyl acetate.

## Characterize the Product

- ❑ To determine the purity of the product, run a TLC of a sample using ethyl acetate as eluent.
- ❑ Obtain the melting point of the product and compare it to the literature value.
- ❑ Obtain an IR spectrum of the product and compare it to that of the starting material.
- ❑ Obtain $^{1}$H-NMR and $^{13}$C-NMR spectra in $CDCl_3$ if instructed to do so.

## QUESTIONS

1. Why is aqueous sodium bicarbonate used in the workup of the ester product? Would sodium hydroxide be an alternative?
2. Write the reaction for the hydrolysis of an ester (the reverse reaction of esterification).
3. Nitriles do not contain a carbonyl group, unlike esters, amides, acid chlorides, and acid anhydrides. Explain why nitriles are still classed as carboxylic acid derivatives.
4. Complete the following reactions that produce essential oils.

(a) O, OH + OH ⟶ *banana-like odor*

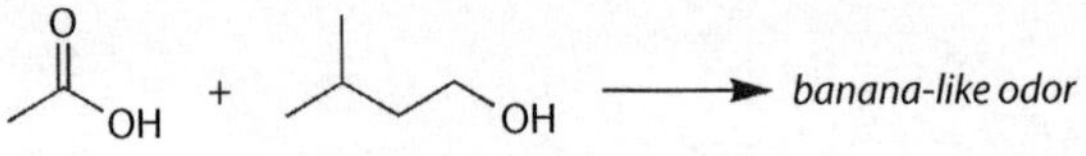

(b) O, OH + OH ⟶ *pineapple-like odor*

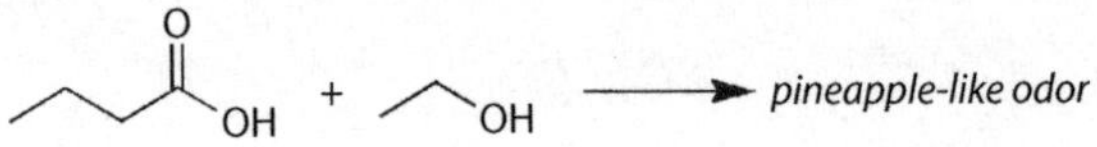

(c) O, OH, OH + $CH_3OH$ ⟶ *wintergreen oil*

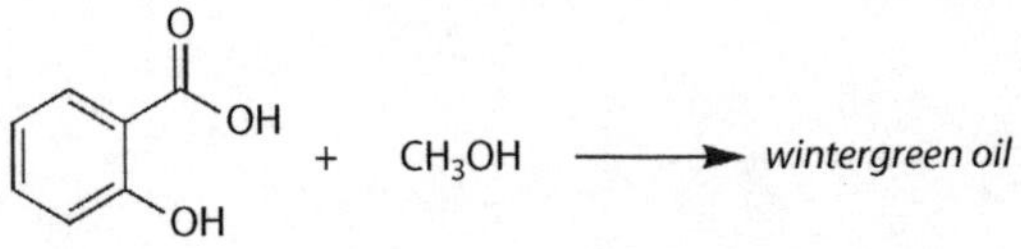

# 5 Transesterification Reaction
## *Preparation of Biodiesel*

**LEARNING GOALS**

- To perform a transesterification reaction
- To determine the outcome of a reaction qualitatively by thin-layer chromatography and quantitatively by nuclear magnetic resonance (NMR) spectroscopy
- To understand the concepts of biofuels production

## INTRODUCTION

At the start of this new millennium, energy demands are ever increasing but fossil fuels are becoming limited. As a result, there is growing interest in developing alternative energy resources. These include hydrogen cells, solar energy, and wind power. However, these technologies are still at the developing stage, and the costs of applying them are high. An immediately applicable option is replacement of diesel fuel by biodiesel, which consists of the simple alkyl esters of fatty acids. With little modification, diesel engine vehicles can use biodiesel fuels. They can also be used as heating oils. Biodiesels are biodegradable and nontoxic and have lower CO and hydrocarbon emissions than petroleum-based diesel when burned. Conversely, they do present other technical challenges such as low cloud points and elevated $NO_x$ emissions.

Biodiesel is generally made from vegetable oils or animal fats by transesterification with methanol. A transesterification is the process of converting one ester into another by exchanging organic groups with an alcohol. The reaction is catalyzed by the addition of an acid or base. The products of the reaction are fatty acid methyl esters (FAME), which are the biodiesel and glycerin, which also have numerous applications, for example, in the food, cosmetic, and pharmaceutical sectors.

*vegetable oil* *methanol* *fatty acid methyl esters (FAME)* *glycerol*

The most commonly used catalysts for transesterification reactions are potassium hydroxide (base catalysis) and sulfuric acid (acid catalysis). In the case of the preparation of biodiesel, the starting oil and methanol are generally heated and stirred with the catalyst, or, in the case of some base catalysts, the catalyst is dissolved in the methanol prior to addition of the oil. There are differences in reaction rate between the acid- and base-catalyzed pathways. These are explored in this experiment: half the group performing the reaction using potassium hydroxide as the catalyst and the other half of the group using sulfuric acid. The mechanism for acid-catalyzed transesterification is:

Both batch and continuous-flow technologies have been used to scale up the reaction. The majority of biodiesel production facilities produce in excess of five million gallons per year using this reaction. This often requires employing large equipment when operating using conventional heating. It is possible to prepare up to seven liters of biodiesel a minute (~110 gallons per hour) using a continuous-flow microwave unit with a small footprint. Four of these units running in parallel could produce around three million gallons of biodiesel per year. As well as requiring much less space than a conventional reactor, another advantage of the microwave approach is that if one of the four units develops a fault, the throughput drops by just 25% until it is fixed, because the other three units can still run. If one large conventional reactor is used and a problem arises, production has to be halted until repairs are complete.

Numerous methods have been developed to assess the degree of conversion of vegetable oil and methanol to biodiesel. Perhaps the most common way to analyze biodiesel is by gas chromatography using the American Society for Testing

Equipment for the continuous-flow preparation of biodiesel

Four units running in parallel:
- 3.4 million gallons of biodiesel per year
- Offers redundancy

[Reproduced with permission from CEM Corporation.]

and Materials (ASTM) test method. For qualitative analysis, thin-layer chromatography can be used. Vegetable oil and biodiesel have significantly different $R_f$ values. The relative size of the spots from starting oil and biodiesel product offers an approximation as to the success of the transesterification reaction.

More quantitatively, $^1$H-NMR spectroscopy can be used. The conversion can be determined based on the integration value of the methyl group on the ester product (at approximately 3.7 ppm) and that of the methylenic group α to the carbonyl functionality in the starting oil and the biodiesel product (at approximately 2.3 ppm). The glycerol by-product separates from the biodiesel before analysis, but also is insoluble in deuterated chloroform. As a result it does not appear in the NMR spectrum. The excess methanol used in the reaction does appear in the NMR spectrum, unless it is removed by evaporation prior to analysis.

*vegetable oil* ⟹ 3 *biodiesel*

*H* = methylenic ($CH_2$) protons α to carbonyl; at ~2.3 ppm in NMR spectrum
Total number of identical protons = 12 (6 from starting material, 6 from product)

H = protons on the methyl ester group ($CH_3$) of biodiesel; at ~3.7 ppm in NMR spectrum
Total number of identical protons = 9

To determine the conversion to biodiesel, a normalization factor has to be applied, due to there being three protons from the methyl ester functionality (ME) and two protons from the methylenic group ($CH_2$). The conversion can be calculated using:

$$\text{Conversion} = 100 \times \frac{2\ I_{ME}}{3\ I_{CH2}}$$

The term $I_{ME}$ is the integration value of the methyl ester protons and the term $I_{CH2}$ is the integration value from the methylenic protons.

# PROCEDURE FOR USE IN A MONOMODE MICROWAVE UNIT

## Preparation of Biodiesel Using Acid Catalysis

$$\text{Triglyceride (R groups)} + 3\ CH_3OH \xrightarrow[H_2SO_4]{MW} 3\ RCO_2CH_3 + HOCH_2CH(OH)CH_2OH$$

**Caution:** The reaction requires the use of concentrated sulfuric acid, which can cause serious burns if spilled on the skin. Transfer the sulfuric acid to a glass graduated cylinder and pipette the required amount directly into the reaction tube. Flush any extra reagent down the sink with large amounts of water. If any sulfuric acid is accidentally spilled on the skin, immediately wash the area with large amounts of water to prevent a burn. Phosphomolybdic acid (PMA) is a powerful oxidant and stain. Do not allow it to come into contact with skin or clothing. The use of goggles with side-shields, lab coats, and gloves is considered minimum and nondiscretionary safety practice in the laboratory.

### Table of Reagents and Physical Constants

| Reagent | Equiv. | FW | mmol | Mass (mg) | Density (g/mL) | Vol. (mL) | mp/bp°C |
|---|---|---|---|---|---|---|---|
| Vegetable oil[a] | 1 | 886 | 2.14 | 1900 | 0.95 | 2.0 | 554 |
| Methanol $CH_4O$ | 11.53 | 32.0 | 24.68 | 790 | 0.79 | 1.0 | 65 |
| Sulfuric acid (conc.) $H_2SO_4$ | 0.87 | 98.1 | 1.87 | 183 | 1.83 | 0.1 | 324 |

[a] Calculations based on vegetable oil having an average molecular weight and density of that of the model compound triolein (9-octadecenoic acid-(Z)-1,2,3-propanetriyl ester).

## Perform the Reaction

- ❑ In a 10-mL glass microwave reaction vessel place a magnetic stir bar, the vegetable oil, and the methanol.
- ❑ In a fume hood carefully add the concentrated sulfuric acid dropwise.
- ❑ Seal the reaction vessel with a cap according to the microwave manufacturer's recommendations.
- ❑ Program the microwave unit to heat the vessel contents to 120°C over a period of 3 minutes and then hold at this temperature for 5 minutes.
- ❑ After the heating step is completed, allow the contents of the reaction vessel to cool to 50°C or below before removing it from the microwave cavity.
- ❑ Allow the reaction mixture to settle. There may be two layers: an upper colorless layer and a yellowish lower layer.

### Determine the Extent of Reaction Using Thin-Layer Chromatography

- ❑ If observed, carefully remove the upper colorless layer from the reaction vessel using a pipette and place it in a test tube.
- ❑ Run a TLC of the remaining yellow material in the reaction vessel by removing a small sample with a micropipette.
- ❑ Spot the material on a prepared TLC plate. Alongside your mixture, spot the starting vegetable oil as a reference.
- ❑ Prepare a TLC chamber by taking a beaker large enough to fit the TLC plate and adding a small layer of 5% ethyl acetate in hexane.
- ❑ Cover the beaker with a watch glass.
- ❑ Carefully place the TLC plate in the beaker, leaning the plate against the side of the beaker.
- ❑ Be sure the spot is above the level of the eluting solvent.
- ❑ Re-cover the beaker.
- ❑ Once the solvent front is near the top of the plate carefully remove the plate from the chamber.
- ❑ Immediately mark the solvent front with a pencil and then allow the plate to dry.
- ❑ Develop the TLC plate using a phosphomolybdic acid stain.
- ❑ Determine the $R_f$ values for the vegetable oil and the biodiesel product.
- ❑ Estimate the extent of the reaction by comparing the relative size of the spots from the starting oil and biodiesel product.

### Quantify the Extent of Reaction Using NMR Spectroscopy

- ❑ With a pipette, place into a test tube two drops of the remaining yellow material in the reaction vessel.
- ❑ With a new clean pipette, add deuterated chloroform ($CDCl_3$,~0.5 mL) to the test tube.
- ❑ Mix the contents of the test tube and then pipette the solution into an NMR tube so that there is a total liquid height of approximately 2.5 cm in the tube.
- ❑ Obtain a $^1$H-NMR spectrum.
- ❑ Integrate the signals in the NMR spectrum at approximately 2.3 ppm and 3.7 ppm. The excess methanol has a singlet at 3.49 ppm. Do not mistake this for the signal for the methyl ester group in biodiesel found at around 3.7 ppm.
- ❑ Using the normalization equation, determine the conversion.

# PROCEDURE FOR USE IN A MULTIMODE MICROWAVE UNIT

## Preparation of Biodiesel Using Acid Catalysis

$$(RCOO)_3C_3H_5 + 3\ CH_3OH \xrightarrow[H_2SO_4]{MW} 3\ RCOOCH_3 + HOCH_2CH(OH)CH_2OH$$

**Caution:** The reaction requires the use of concentrated sulfuric acid, which can cause serious burns if spilled on the skin. Transfer the sulfuric acid to a glass graduated cylinder and pipette the required amount directly into the reaction tube. Flush any extra reagent down the sink with large amounts of water. If any sulfuric acid is accidentally spilled on the skin, immediately wash the area with large amounts of water to prevent a burn. Phosphomolybdic acid is a powerful oxidant and stain. Do not allow it to come into contact with skin or clothing. The use of goggles with side-shields, lab coats, and gloves is considered minimum and nondiscretionary safety practice in the laboratory.

---

### Table of Reagents and Physical Constants

| Reagent | Equiv. | FW | mmol | Mass (mg) | Density (g/mL) | Vol. (mL) | mp/bp°C |
|---|---|---|---|---|---|---|---|
| Vegetable oil[a] | 1 | 886 | 4.28 | 3800 | 0.95 | 4.0 | 554 |
| Methanol $CH_4O$ | 11.53 | 32.0 | 49.38 | 158 | 0.79 | 2.0 | 65 |
| Sulfuric acid (conc.) $H_2SO_4$ | 0.87 | 98.1 | 3.74 | 366 | 1.83 | 0.2 | 324 |

[a] Calculations based on vegetable oil having an average molecular weight and density of that of the model compound triolein (9-octadecenoic acid-(*Z*)-1,2,3-propanetriyl ester).

---

## Perform the Reaction

- ❑ In a 25-mL glass microwave reaction vessel place a magnetic stir bar, the vegetable oil, and the methanol.
- ❑ In a fume hood carefully add the concentrated sulfuric acid dropwise.
- ❑ Seal the reaction vessel with a cap according to the microwave manufacturer's recommendations.
- ❑ Place the sealed reaction vessel in the carousel, noting the vessel's position number and ensuring that vessels are evenly spaced around the carousel.
- ❑ When all the group's reaction vessels are in place, load the carousel into the microwave cavity.
- ❑ If provided by the manufacturer, connect the temperature probe to the control vessel.

- ❑ Program the microwave unit to heat the vessel contents to 120°C over a period of 10 minutes and then hold at this temperature for 5 minutes.
- ❑ After the heating step is completed, allow the contents of the reaction vessel to cool to 50°C or below before removing it from the microwave cavity.
- ❑ Allow the reaction mixture to settle. There may be two layers: an upper colorless layer and a yellowish lower layer.

## Determine the Extent of Reaction Using Thin-Layer Chromatography

- ❑ If observed, carefully remove the upper colorless layer from the reaction vessel using a pipette and place it in a test tube.
- ❑ Run a TLC of the remaining yellow material in the reaction vessel by removing a small sample with a micropipette.
- ❑ Spot the material on a prepared TLC plate. Alongside your mixture, spot the starting vegetable oil as a reference.
- ❑ Prepare a TLC chamber by taking a beaker large enough to fit the TLC plate and adding a small layer of 5% ethyl acetate in hexane.
- ❑ Cover the beaker with a watch glass.
- ❑ Carefully place the TLC plate in the beaker leaning the plate against the side of the beaker.
- ❑ Be sure the spot is above the level of the eluting solvent.
- ❑ Re-cover the beaker.
- ❑ Once the solvent front is near the top of the plate carefully remove the plate from the chamber.
- ❑ Immediately mark the solvent front with a pencil and then allow the plate to dry.
- ❑ Develop the TLC plate using a phosphomolybdic acid stain.
- ❑ Determine the $R_f$ values for the vegetable oil and the biodiesel product.
- ❑ Estimate the extent of the reaction by comparing the relative size of the spots from the starting oil and biodiesel product.

## Quantify the Extent of Reaction Using NMR Spectroscopy

- ❑ With a pipette, place into a test tube two drops of the remaining yellow material in the reaction vessel.
- ❑ With a new clean pipette, add deuterated chloroform ($CDCl_3$, ~0.5 mL) to the test tube.
- ❑ Mix the contents of the test tube and then pipette the solution into an NMR tube so that there is a total liquid height of approximately 2.5 cm in the tube.
- ❑ Obtain a $^1$H-NMR spectrum.
- ❑ Integrate the signals in the NMR spectrum at approximately 2.3 ppm and 3.7 ppm. The excess methanol has a singlet at 3.49 ppm. Do not mistake this for the signal for the methyl ester group in biodiesel found at around 3.7 ppm.
- ❑ Using the normalization equation, determine the conversion.

# PROCEDURE FOR USE IN A MONOMODE MICROWAVE UNIT

## PREPARATION OF BIODIESEL USING BASE CATALYSIS

triglyceride + 3 $CH_3OH$ —(MW, KOH)→ 3 $RCO_2CH_3$ + $HOCH_2CH(OH)CH_2OH$

**Caution:** The reaction requires the use of potassium hydroxide, a strong base, which can cause serious burns if in contact with the skin. Measure potassium hydroxide with extreme care. If any solid base or base in solvent accidentally comes into contact with the skin, immediately wash the area with large amounts of water to prevent a burn. The use of goggles with side-shields, lab coats, and gloves is considered minimum and nondiscretionary safety practice in the laboratory.

### Table of Reagents and Physical Constants

| Reagent | Equiv. | FW | mmol | Mass (mg) | Density (g/mL) | Vol. (mL) | mp/bp°C |
|---|---|---|---|---|---|---|---|
| Vegetable oil[a] | 1 | 886 | 2.14 | 1900 | 0.95 | 2.0 | 554 |
| Methanol $CH_4O$ | 11.53 | 32.0 | 24.68 | 790 | 0.79 | 1.0 | 65 |
| Potassium hydroxide NaOH | 1.78 | 56.1 | 1.87 | 100 | — | — | — |

[a] Calculations based on vegetable oil having an average molecular weight and density of that of the model compound triolein (9-octadecenoic acid-(Z)-1,2,3-propanetriyl ester).

## PERFORM THE REACTION

- ❑ In a 10-mL glass microwave reaction vessel place a magnetic stir bar and the vegetable oil.
- ❑ Place the potassium hydroxide in a 10-mL Erlenmeyer flask.
- ❑ Carefully add the methanol to the flask containing the potassium hydroxide and swirl or stir with a glass rod to dissolve the base.
- ❑ Add 1 mL of the potassium hydroxide in methanol solution to the reaction vessel.
- ❑ Seal the reaction vessel with a cap according to the microwave manufacturer's recommendations.
- ❑ Program the microwave unit to heat the vessel contents to 120°C over a period of 3 minutes and then hold at this temperature for 5 minutes.
- ❑ After the heating step is completed, allow the contents of the reaction vessel to cool to 50°C or below before removing it from the microwave cavity.

- ❑ Remove the cap from the reaction vessel and add 2 M hydrochloric acid (2 mL).
- ❑ Clamp the vessel over a stirring hotplate and stir the contents for 2 minutes. This quenches the reaction.
- ❑ Allow the reaction mixture to settle. There may be two layers: an upper colorless layer and a yellowish lower layer.

## Determine the Extent of Reaction Using Thin-Layer Chromatography

- ❑ If observed, carefully remove the upper colorless layer from the reaction vessel using a pipette and place it in a test tube.
- ❑ Run a TLC of the remaining yellow material in the reaction vessel by removing a small sample with a micropipette.
- ❑ Spot the material on a prepared TLC plate. Alongside the mixture, spot the starting vegetable oil as a reference.
- ❑ Prepare a TLC chamber by taking a beaker large enough to fit the TLC plate and adding a small layer of 5% ethyl acetate in hexane.
- ❑ Cover the beaker with a watch glass.
- ❑ Carefully place the TLC plate in the beaker leaning the plate against the side of the beaker.
- ❑ Be sure the spot is above the level of the eluting solvent.
- ❑ Re-cover the beaker.
- ❑ Once the solvent front is near the top of the plate carefully remove the plate from the chamber.
- ❑ Immediately mark the solvent front with a pencil and then allow the plate to dry.
- ❑ Develop the TLC plate using a phosphomolybdic acid stain.
- ❑ Determine the $R_f$ values for the vegetable oil and the biodiesel product.
- ❑ Estimate the extent of the reaction by comparing the relative size of the spots from the starting oil and biodiesel product.

## Quantify the Extent of Reaction Using NMR Spectroscopy

- ❑ With a pipette, place into a test tube two drops of the remaining yellow material in the reaction vessel.
- ❑ With a new clean pipette, add deuterated chloroform ($CDCl_3$,~0.5 mL) to the test tube.
- ❑ Mix the contents of the test tube and then pipette the solution into an NMR tube so that there is a total liquid height of approximately 2.5 cm in the tube.
- ❑ Obtain a $^1$H-NMR spectrum.
- ❑ Integrate the signals in the NMR spectrum at approximately 2.3 ppm and 3.7 ppm. The excess methanol has a singlet at 3.49 ppm. Do not mistake this for the signal for the methyl ester group in biodiesel found at around 3.7 ppm.
- ❑ Using the normalization equation, determine the conversion.

# PROCEDURE FOR USE IN A MULTIMODE MICROWAVE UNIT

## Preparation of Biodiesel Using Base Catalysis

$$RCOOCH_2CH(OCOR)CH_2OCOR + 3\ CH_3OH \xrightarrow[\text{KOH}]{\text{MW}} 3\ RCOOCH_3 + HOCH_2CH(OH)CH_2OH$$

**Caution:** The reaction requires the use of potassium hydroxide, a strong base, which can cause serious burns if in contact with the skin. Measure potassium hydroxide with extreme care. If any solid base or base in solvent accidentally comes into contact with the skin, immediately wash the area with large amounts of water to prevent a burn. The use of goggles with side-shields, lab coats, and gloves is considered minimum and nondiscretionary safety practice in the laboratory.

---

### Table of Reagents and Physical Constants

| Reagent | Equiv. | FW | mmol | Mass (g) | Density (g/mL) | Vol. (mL) | mp/bp°C |
|---|---|---|---|---|---|---|---|
| Vegetable oil[a] | 1 | 886 | 4.28 | 3.80 | 0.95 | 4.0 | 554 |
| Methanol | 57.68 | 32.0 | 123.44 | 3.95 | 0.79 | 5.0 | 65 |
| $CH_4O$ | | | | | | | |
| Potassium hydroxide | 1.78 | 56.1 | 1.87 | 0.10 | — | — | — |
| NaOH | | | | | | | |

[a] Calculations based on vegetable oil having an average molecular weight and density of that of the model compound triolein (9-octadecenoic acid-(Z)-1,2,3-propanetriyl ester).

---

## Perform the Reaction

- ❑ In a 25-mL glass microwave reaction vessel place a magnetic stir bar and the vegetable oil.
- ❑ Place the potassium hydroxide in a 10-mL Erlenmeyer flask.
- ❑ Carefully add the methanol to the flask containing the potassium hydroxide and swirl or stir with a glass rod to dissolve the base.
- ❑ Add 1 mL of the potassium hydroxide in methanol solution to the reaction vessel.
- ❑ Seal the reaction vessel with a cap according to the microwave manufacturer's recommendations.
- ❑ Place the sealed reaction vessel in the carousel, noting the vessel's position number and ensuring that vessels are evenly spaced around the carousel.
- ❑ When all the group's reaction vessels are in place, load the carousel into the microwave cavity.
- ❑ If provided by the manufacturer, connect the temperature probe to the control vessel.

- ❑ Program the microwave unit to heat the vessel contents to 120°C over a period of 3 minutes and then hold at this temperature for 5 minutes.
- ❑ After the heating step is completed, allow the contents of the reaction vessel to cool to 50°C or below before removing it from the microwave cavity.
- ❑ Remove the cap from the reaction vessel and add 2 M hydrochloric acid (2 mL).
- ❑ Clamp the vessel over a stirring hotplate and stir the contents for 2 minutes. This quenches the reaction.
- ❑ Allow the reaction mixture to settle. There may be two layers: an upper colorless layer and a yellowish lower layer.

## Determine the Extent of Reaction Using Thin-Layer Chromatography

- ❑ If observed, carefully remove the upper colorless layer from the reaction vessel using a pipette and place it in a test tube.
- ❑ Run a TLC of the remaining yellow material in the reaction vessel by removing a small sample with a micropipette.
- ❑ Spot the material on a prepared TLC plate. Alongside the mixture, spot the starting vegetable oil as a reference.
- ❑ Prepare a TLC chamber by taking a beaker large enough to fit the TLC plate and adding a small layer of 5% ethyl acetate in hexane.
- ❑ Cover the beaker with a watch glass.
- ❑ Carefully place the TLC plate in the beaker leaning the plate against the side of the beaker.
- ❑ Be sure the spot is above the level of the eluting solvent.
- ❑ Re-cover the beaker.
- ❑ Once the solvent front is near the top of the plate carefully remove the plate from the chamber.
- ❑ Immediately mark the solvent front with a pencil and then allow the plate to dry.
- ❑ Develop the TLC plate using a phosphomolybdic acid stain.
- ❑ Determine the $R_f$ values for the vegetable oil and the biodiesel product.
- ❑ Estimate the extent of the reaction by comparing the relative size of the spots from the starting oil and biodiesel product.

## Quantify the Extent of Reaction Using NMR Spectroscopy

- ❑ With a pipette, place into a test tube two drops of the remaining yellow material in the reaction vessel.
- ❑ With a new clean pipette, add deuterated chloroform ($CDCl_3$, ~0.5 mL) to the test tube.
- ❑ Mix the contents of the test tube and then pipette the solution into an NMR tube so that there is a total liquid height of approximately 2.5 cm in the tube.
- ❑ Obtain a $^1$H-NMR spectrum.

- ❑ Integrate the signals in the NMR spectrum at approximately 2.3 ppm and 3.7 ppm. The excess methanol has a singlet at 3.49 ppm. Do not mistake this for the signal for the methyl ester group in biodiesel found at around 3.7 ppm.
- ❑ Using the normalization equation, determine the conversion.

## QUESTIONS

1. Draw the mechanism for the following base-catalyzed transesterification reaction.

$$R{-}C({=}O){-}O{-}R^1 + H{-}O{-}R^2 \xrightleftharpoons{KOH} R{-}C({=}O){-}O{-}R^2 + H{-}O{-}R^1$$

2. A difference in rate between the acid and base-catalyzed transesterification reactions is observed in the preparation of biodiesel. What is the origin of these differences?
3. The biofuels industry has been criticized over the last few years. What is the current thinking on the viability of biofuels such as biodiesel and bioethanol? Do you agree with the criticisms?

# 6 Knoevenagel Condensation Reaction

## *Preparation of 3-Acetylcoumarin*

**LEARNING GOALS**

- To perform a condensation reaction
- To prepare a heterocyclic compound

## INTRODUCTION

Heterocyclic compounds constitute the largest and most varied family of organic compounds. Heterocycles are defined as cyclic compounds that have atoms of at least two different elements in the ring(s). Coumarins are one class in the heterocyclic family. The parent compound, coumarin, is found in many plants. It is known for its sweet smell of fresh-cut hay and sweet clover and thus is used as a component in perfumes. Derivatives of coumarin are also found in the pharmaceutical industry, acting as anticoagulants (blood thinners). In this experiment we synthesize 3-acetylcoumarin, by the condensation of salicylaldehyde and ethyl acetoacetate in the presence of piperidine that acts as a base catalyst.

*coumarin*

*salicylaldehyde* + *ethyl acetoacetate* ($OCH_2CH_3$) —(*piperidine*)→ *3-acetylcoumarin* + $CH_3CH_2OH$

This synthetic approach to 3-acetylcoumarin involves two key chemical transformations: a transesterification and a Knoevenagel condensation. A transesterification is the process of converting one ester into another by exchanging organic groups with an alcohol. The reaction is catalyzed by the addition of an acid or base.

$$RC(=O)OR' + HOR'' \xrightarrow{\text{catalyst}} RC(=O)OR'' + HOR'$$

A condensation reaction is when two molecules or functional groups combine to make a single compound with the loss of a small molecule, usually water. In the formation of peptides from amino acids, the condensation is between an amine and a carboxylic acid to form the amide functional group plus water. In the case of an aldol condensation the reaction is between an enolate anion and the carbonyl of an aldehyde to form β-hydroxy carbonyls that then lose a molecule of water to form α,β-unsaturated aldehydes. In the Knoevenagel condensation, the functional groups involved in the condensation are an aldehyde and an activated $-CH_2-$ group (called a methylene) that is deprotonated by an amine base, the product of the reaction being an α,β unsaturated compound and water.

$$H_2NCH(R^1)COOH + H_2NCH(R^2)CO_2H \xrightarrow{\text{coupling agents}} H_2NCH(R^1)C(=O)NHCH(R^2)CO_2H + HOH$$

*Peptide Condensation*

$$R^1C(=O)CH_2R^2 + R^3C(=O)R^4 \xrightarrow{\text{base}} R^1C(=O)C(R^2)=C(R^3)R^4 + H_2O$$

*Aldol Condensation*

$$CH_2R^1R^2 + R^3C(=O)R^4 \xrightarrow{\text{base}} R^1(R^2)C=C(R^3)R^4 + H_2O$$

*Knoevenagel Condensation*

Looking at the structure of 3-acetylcoumarin, the transesterification reaction forms the carbon–oxygen single bond in the heterocyclic ring and the Knoevenagel condensation forms a carbon–carbon double bond.

*Knoevenagel condensation*

*transesterification*

The order in which the condensation and transesterification take place in the reaction of salicylaldehyde with ethyl acetoacetate is not fully known. A model reaction of benzaldehyde with ethyl acetoacetate does illustrate the mechanism of the Knoevenagel condensation. The base catalyst first deprotonates the ethyl acetoacetate to form a resonance-stabilized anion.

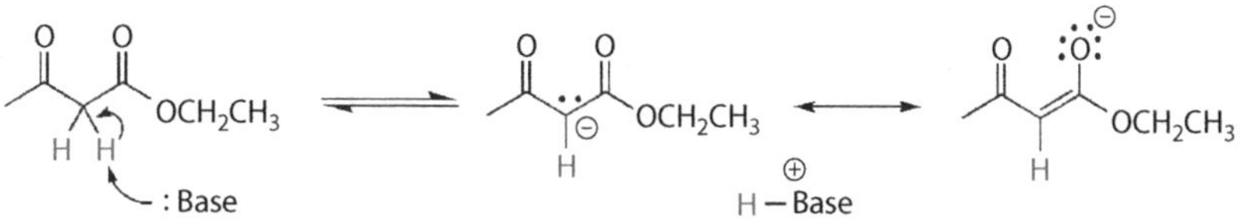

The anion then reacts with benzaldehyde.

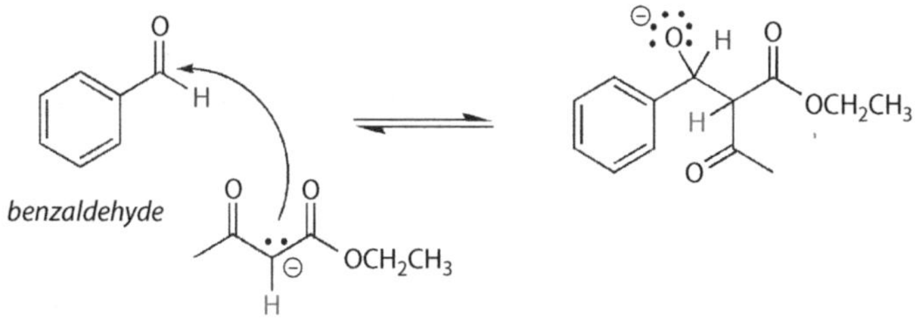

The product is formed after protonation of the oxygen anion followed by a base-induced elimination.

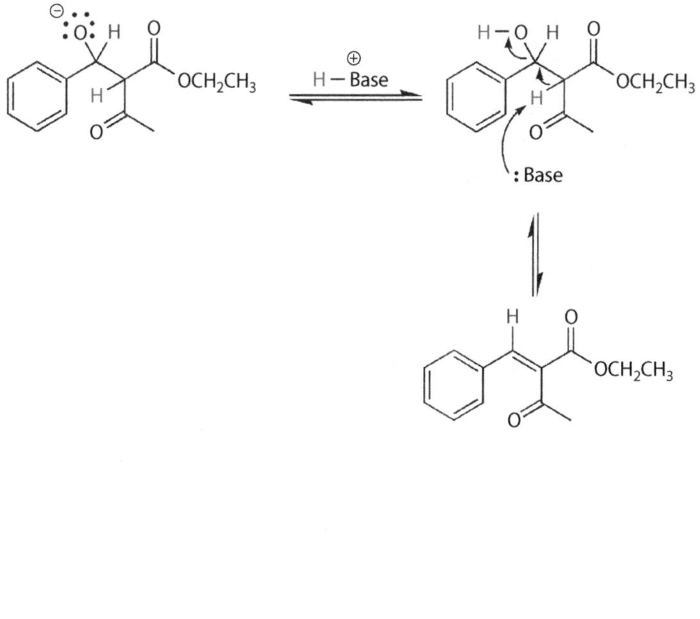

## PROCEDURE FOR USE IN A MONOMODE MICROWAVE UNIT

O OH + O O $OCH_2CH_3$ —MW, piperidine→ O O O + $CH_3CH_2OH$

**Caution:** Salicylaldehyde, ethyl acetoacetate, and piperidine are irritants. Piperidine is also toxic and should be dispensed in a hood. The use of goggles with side-shields, lab coats, and gloves is considered minimum and nondiscretionary safety practice in the laboratory.

### Table of Reagents and Physical Constants

| Reagent | Equiv. | FW | mmol | Mass (mg) | Density (g/mL) | Vol. (mL) | mp/bp°C |
|---|---|---|---|---|---|---|---|
| Salicylaldehyde<br>$C_7H_6O_2$ | 1 | 122 | 3 | 366 | 1.16 | 0.31 | 197 |
| Ethyl acetoacetate<br>$C_6H_{10}O_3$ | 1 | 130 | 3 | 390 | 1.03 | 0.38 | 181 |
| Piperidine<br>$C_5H_{11}N$ | 0.1 | 85.0 | 0.3 | 26 | 0.862 | 0.030 | 106 |
| Ethyl acetate<br>$C_4H_8O_2$ | — | — | — | — | — | 1.0 | 77 |

### PERFORM THE REACTION

- ❑ In a 10-mL glass microwave reaction vessel, place a magnetic stir bar.
- ❑ Using an automatic delivery pipette add the salicylaldehyde, ethyl acetoacetate, and piperidine to the reaction vessel.
- ❑ Using a graduated cylinder add the ethyl acetate to the reaction vessel.
- ❑ Seal the reaction vessel with a cap according to the microwave manufacturer's recommendations.
- ❑ Carefully stir the contents on a magnetic stir plate to dissolve the solid.
- ❑ Place the sealed reaction vessel into the microwave cavity.
- ❑ Program the microwave unit to heat the vessel contents to 130°C over a 2-minute ramp period and then hold at this temperature for 8 minutes.
- ❑ After the heating step is completed, allow the contents of the reaction vessel to cool to 50°C or below before removing it from the microwave cavity.

### ISOLATE THE PRODUCT

- ❑ Prepare two ice baths by taking crushed ice and placing it in two beakers.
- ❑ Carefully open the reaction vessel.

- ❑ Remove the stirring bar with a magnetic retrieving wand or a pair of tweezers.
- ❑ Slowly add hexanes (4 mL).
- ❑ Cool the solution in one of the ice baths for 10 minutes after the addition is complete.
- ❑ While the solution is cooling set up a vacuum filtration system with a Hirsch funnel, side-arm flask, rubber collar, and a length of rubber vacuum tubing.
- ❑ In the other ice bath, cool some hexanes (2–4 mL) in a 25-mL Erlenmeyer flask for washing the precipitate after filtration.
- ❑ Connect the filtration system to a vacuum and place the correct size filter paper in the funnel.
- ❑ Wet the filter paper with a few drops of the cold hexanes and start the vacuum to seal the filter paper in place.
- ❑ Filter the reaction mixture by pouring the contents into the funnel; transfer as much solid as possible.
- ❑ Rinse the reaction vessel with cold distilled hexanes (1 mL) and add the washings to the filter funnel.
- ❑ Rinse the precipitate on the filter with additional cold hexanes (1 mL).
- ❑ Allow the precipitate to dry on the funnel for 10 minutes.
- ❑ Transfer the solid to a large piece of filter paper to dry completely.
- ❑ Weigh the dry product and calculate the yield and percentage yield.

## Characterize the Product

- ❑ Determine the melting point of the product.
- ❑ If the melting point indicates the product is not pure, re-crystallize the crude product from ethanol.
- ❑ Re-calculate the percent yield.
- ❑ Obtain an IR spectrum of the product and compare it with that of the starting material.
- ❑ Obtain $^{1}$H-NMR and $^{13}$C-NMR spectra in $CDCl_3$ if instructed to do so.

## PROCEDURE FOR USE IN A MULTIMODE MICROWAVE UNIT

OH + $OCH_2CH_3$ → (MW, piperidine) + $CH_3CH_2OH$

**Caution:** Salicylaldehyde, ethyl acetoacetate, and piperidine are irritants. Piperidine is also toxic and should be dispensed in a hood. The use of goggles with side-shields, lab coats, and gloves is considered minimum and nondiscretionary safety practice in the laboratory.

### Table of Reagents and Physical Constants

| Reagent | Equiv. | FW | mmol | Mass (mg) | Density (g/mL) | Vol. (mL) | mp/bp°C |
|---|---|---|---|---|---|---|---|
| Salicylaldehyde $C_7H_6O_2$ | 1 | 122 | 3 | 366 | 1.16 | 0.31 | 197 |
| Ethyl acetoacetate $C_6H_{10}O_3$ | 1 | 130 | 3 | 390 | 1.03 | 0.38 | 181 |
| Piperidine $C_5H_{11}N$ | 0.1 | 85.0 | 0.3 | 26 | 0.862 | 0.030 | 106 |
| Ethyl acetate $C_4H_8O_2$ | — | — | — | — | — | 4.0 | 77 |

### Perform the Reaction

- ❑ In a 25-mL glass microwave reaction vessel, place a magnetic stir bar.
- ❑ Using an automatic delivery pipette add the salicylaldehyde, ethyl acetoacetate, and piperidine to the reaction vessel.
- ❑ Using a graduated cylinder add the ethyl acetate to the reaction vessel.
- ❑ Seal the reaction vessel with a cap according to the microwave manufacturer's recommendations.
- ❑ Carefully stir the contents on a magnetic stir plate to dissolve the solid.
- ❑ Place the sealed reaction vessel in the carousel, noting the vessel's position number and ensuring that vessels are evenly spaced around the carousel.
- ❑ When all the group's reaction vessels are in place, load the carousel into the microwave cavity.
- ❑ If provided by the manufacturer, connect the temperature probe to the control vessel.
- ❑ Program the microwave unit to heat the vessel contents to 130°C over a 2-minute ramp period and then hold at this temperature for 8 minutes.
- ❑ After the heating step is completed, allow the contents of the reaction vessel to cool to 50°C or below before removing it from the microwave cavity.

## Isolate the Product

- ❑ Prepare two ice baths by taking crushed ice and placing it in two beakers.
- ❑ Carefully open the reaction vessel.
- ❑ Remove the stirring bar with a magnetic retrieving wand or a pair of tweezers.
- ❑ Slowly add hexanes (4 mL).
- ❑ Cool the solution in one of the ice baths for 10 minutes after the addition is complete.
- ❑ While the solution is cooling set up a vacuum filtration system with a Hirsch funnel, side-arm flask, rubber collar, and a length of rubber vacuum tubing.
- ❑ In the other ice bath, cool some hexanes (2–4 mL) in a 25-mL Erlenmeyer flask for washing the precipitate after filtration.
- ❑ Connect the filtration system to a vacuum and place the correct size filter paper in the funnel.
- ❑ Wet the filter paper with a few drops of the cold hexanes and start the vacuum to seal the filter paper in place.
- ❑ Filter the reaction mixture by pouring the contents into the funnel; transfer as much solid as possible.
- ❑ Rinse the reaction vessel with cold distilled hexanes (1 mL) and add the washings to the filter funnel.
- ❑ Rinse the precipitate on the filter with additional cold hexanes (1 mL).
- ❑ Allow the precipitate to dry on the funnel for 10 minutes.
- ❑ Transfer the solid to a large piece of filter paper to dry completely.
- ❑ Weigh the dry product and calculate the yield and percentage yield.

## Characterize the Product

- ❑ Determine the melting point of the product.
- ❑ If the melting point indicates the product is not pure, re-crystallize the crude product from ethanol.
- ❑ Re-calculate the percent yield.
- ❑ Obtain an IR spectrum of the product and compare it with that of the starting material.
- ❑ Obtain $^{1}H$-NMR and $^{13}C$-NMR spectra in $CDCl_3$ if instructed to do so.

## QUESTIONS

1. For the following reaction, identify the Lewis acid and Lewis base for both the forward and backward reactions. Does the equilibrium favor the starting materials or products?

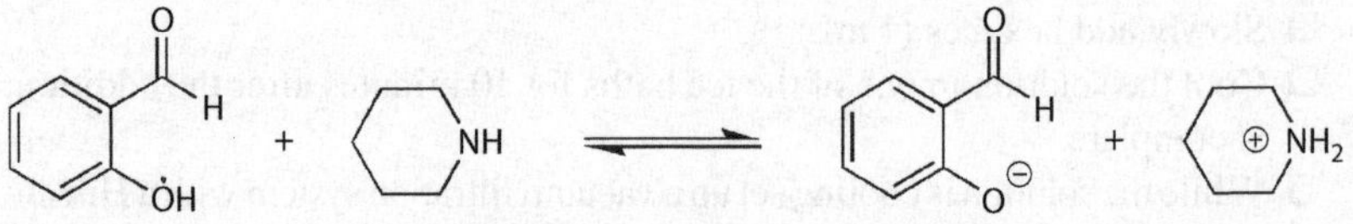

2. Draw the structure of 4-hydroxycoumarin, a biologically active blood thinner.
3. How can you tell from the IR spectrum whether the isolated material in the synthesis of 3-acetylcoumarin is product or one of the starting materials?
4. There are two possible mechanisms for the reaction of salicylaldehyde and ethyl acetoacetate to yield 3-acetylcoumarin, depending on whether the Knoevenagel condensation step takes place before or after the transesterification step. Draw the two mechanisms.

# 7 The Perkin Reaction

## *Condensation of an Aromatic Aldehyde with Rhodanine**

**LEARNING GOALS**

- To perform a condensation reaction involving a carbonyl compound
- To separate the product from the reactants using crystallization and vacuum filtration

## INTRODUCTION

The condensation of a carbonyl compound with an activated $-CH_2$- group (called a methylene) to form a carbon–carbon bond is an important synthetic method. The carbonyl compound can be an aldehyde, ketone, or ester where the carbon of the functional group bears a partially positive charge. The methylene-containing compound acts as a nucleophile, being activated by an electron-withdrawing group such as a carbonyl or nitro group. Examples of these reactions taught in organic chemistry are the aldol condensation, Claisen condensation, Dieckmann condensation, and Perkin reaction.

The Perkin reaction can be used to make cinnamic acids ($\alpha$-$\beta$-unsaturated aromatic acids). It employs an acid anhydride such as acetic anhydride as the activated methylene compound to react with aromatic aldehydes such as benzaldehyde in the presence of a base catalyst.

H + O O O → NaOAc / *acid workup* → OH

*benzaldehyde* *acetic anhydride* *NaOAc = sodium acetate* *cinnamic acid*

* Modified from a conventional procedure: D. W. Mayo, R. M. Pike, and D. C. Forbes, *Microscale Organic Laboratory* (5th ed.), Wiley, New York, 2010, pp. 292–293.

In this experiment the heterocycle, rhodanine, is used as the activated methylene compound instead of an anhydride. Deprotonation of the rhodamine by sodium acetate produces a reactive carbanion intermediate. Attack on the carbonyl group of 2-chlorobenzaldehyde by the carbanion forms a sigma carbon–carbon bond and a β-hydroxy carbonyl after protonation. Under the conditions of this reaction, the β-hydroxy carbonyl intermediate then undergoes a dehydration to form an α-β-unsaturated product.

*rhodanine*

*2-chlorobenzaldehyde*

$H^+$

$-H_2O$

*α, β-unsaturated product*

*β-hydroxy carbonyl compound*

## PROCEDURE FOR USE IN A MONOMODE MICROWAVE UNIT

### Reaction of Rhodanine with 2-Chlorobenzaldehyde

MW
NaOAc
+ $H_2O$

**Caution:** 2-Chlorobenzaldehyde and acetic acid are corrosive. Rhodanine is an irritant. The use of goggles with side-shields, lab coats, and gloves is considered minimum and nondiscretionary safety practice in the laboratory.

### Table of Reagents and Physical Constants

| Reagent | Equiv. | FW | mmol | Mass (mg) | Density (g/mL) | Vol. (mL) | mp/bp°C |
|---|---|---|---|---|---|---|---|
| 2-Chlorobenzaldehyde $C_7H_5ClO$ | 1.8 | 141 | 0.41 | 58 | 1.25 | 0.047 | 212 |
| Rhodanine $C_3H_3NOS_2$ | 1 | 133 | 0.23 | 31 | — | — | 170 |
| Anhydrous sodium acetate (dried)[a] $C_2H_3O_2Na$ | 2.7 | 82.0 | 0.63 | 52 | — | — | 324 |
| Glacial acetic acid $C_2H_4O_2$ | — | — | — | — | — | 1.0 | 118 |

[a] Dry the sodium acetate in an oven for at least one hour before using.

### Perform the Reaction

- ❑ In a 10-mL glass microwave reaction vessel place a magnetic stir bar, the rhodanine, and the sodium acetate.
- ❑ In a fume hood add the glacial acetic acid using a graduated cylinder.
- ❑ Using an automatic delivery pipette, add the 2-chlorobenzaldehyde to the reaction vessel.
- ❑ Seal the reaction vessel with a cap according to the microwave manufacturer's recommendations.
- ❑ Carefully stir the contents on a magnetic stir plate for 1 minute.
- ❑ Place the sealed reaction vessel into the microwave cavity.
- ❑ Program the microwave unit to heat the vessel contents to 180°C over a 1-minute ramp period and then hold at this temperature for 4 minutes.
- ❑ After the heating step is completed, allow the contents of the reaction vessel to cool to 50°C or below before removing it from the microwave cavity.

## Isolate the Product

- ❑ In a fume hood, cool the reaction vessel and its contents in an ice bath for 10 minutes over which time a solid will precipitate from the solution.
- ❑ While the solution is cooling set up a vacuum filtration system with a Hirsch funnel, side-arm flask, rubber collar, and a length of rubber vacuum tubing.
- ❑ In an ice bath, cool some glacial acetic acid (5 mL) in a 25-mL Erlenmeyer flask for washing the precipitate after filtration.
- ❑ Connect the filtration system to a vacuum and place the correct size filter paper in the funnel.
- ❑ Wet the filter paper with a few drops of the cold glacial acetic acid and start the vacuum to seal the filter paper in place.
- ❑ Carefully open the reaction vessel.
- ❑ Filter the reaction mixture by pouring the contents into the funnel; transfer as much solid as possible.
- ❑ Rinse the reaction vessel with cold glacial acetic acid (1 mL) and add the washings to the filter funnel.
- ❑ Rinse the precipitate on the filter with additional glacial acetic acid (1 mL).
- ❑ Allow the precipitate to dry on the funnel for 5 minutes.
- ❑ Transfer the yellow solid to a larger piece of filter paper to complete drying.
- ❑ Weigh the dried product and calculate the yield and percent yield.

## Characterize the Product

- ❑ Determine the melting point of the product and compare it to the literature value.
- ❑ If the melting point is low the product can be re-crystallized from glacial acetic acid.
- ❑ Obtain an IR spectrum of the product and compare it with that of the starting material.
- ❑ Obtain $^{1}$H-NMR and $^{13}$C-NMR spectra in $d_6$-DMSO if instructed to do so.

# PROCEDURE FOR USE IN A MULTIMODE MICROWAVE UNIT

## Reaction of Rhodanine with 2-Chlorobenzaldehyde

MW
NaOAc
+ $H_2O$

**Caution:** 2-Chlorobenzaldehyde and acetic acid are corrosive. Rhodanine is an irritant. The use of goggles with side-shields, lab coats, and gloves is considered minimum and nondiscretionary safety practice in the laboratory.

---

### Table of Reagents and Physical Constants

| Reagent | Equiv. | FW | mmol | Mass (mg) | Density (g/mL) | Vol. (mL) | mp/bp°C |
|---|---|---|---|---|---|---|---|
| 2-Chlorobenzaldehyde $C_7H_5ClO$ | 1.8 | 141 | 0.81 | 114 | 1.25 | 0.091 | 212 |
| Rhodanine $C_3H_3NOS_2$ | 1 | 133 | 0.45 | 60 | — | — | 170 |
| Anhydrous sodium acetate (dried)[a] $C_2H_3O_2Na$ | 2.7 | 82.0 | 1.21 | 100 | — | — | 324 |
| Glacial acetic acid $C_2H_4O_2$ | — | — | — | — | — | 4.0 | 118 |

[a] Dry the sodium acetate in an oven for at least one hour before using.

---

## Perform the Reaction

- ❑ In a 25-mL glass microwave reaction vessel place a magnetic stir bar, the rhodanine, and the sodium acetate.
- ❑ In a fume hood add the glacial acetic acid using a graduated cylinder.
- ❑ Using an automatic delivery pipette, add the 2-chlorobenzaldehyde to the reaction vessel.
- ❑ Seal the reaction vessel with a cap according to the microwave manufacturer's recommendations.
- ❑ Place the sealed reaction vessel in the carousel, noting the vessel's position number and ensuring that vessels are evenly spaced around the carousel.
- ❑ When all the group's reaction vessels are in place, load the carousel into the microwave cavity.
- ❑ If provided by the manufacturer, connect the temperature probe to the control vessel.

- ❑ Program the microwave unit to heat the vessel contents to 180°C over a 2-minute ramp period and then hold at this temperature for 4 minutes.
- ❑ After the heating step is completed, allow the contents of the reaction vessel to cool to 50°C or below before removing it from the microwave cavity.

## Isolate the Product

- ❑ In a fume hood, cool the reaction vessel and its contents in an ice bath for 10 minutes over which time a solid will precipitate from the solution.
- ❑ While the solution is cooling set up a vacuum filtration system with a Hirsch funnel, side-arm flask, rubber collar, and a length of rubber vacuum tubing.
- ❑ In an ice bath, cool some glacial acetic acid (10 mL) in a 25-mL Erlenmeyer flask for washing the precipitate after filtration.
- ❑ Connect the filtration system to a vacuum and place the correct size filter paper in the funnel.
- ❑ Wet the filter paper with a few drops of the cold glacial acetic acid and start the vacuum to seal the filter paper in place.
- ❑ Carefully open the reaction vessel.
- ❑ Filter the reaction mixture by pouring the contents into the funnel; transfer as much solid as possible.
- ❑ Rinse the reaction vessel with cold glacial acetic acid (2 mL) and add the washings to the filter funnel.
- ❑ Rinse the precipitate on the filter with additional glacial acetic acid (2 mL).
- ❑ Allow the precipitate to dry on the funnel for 5 minutes.
- ❑ Transfer the yellow solid to a larger piece of filter paper to complete drying.
- ❑ Weigh the dried product and calculate the yield and percent yield.

## Characterize the Product

- ❑ Determine the melting point of the product and compare it to the literature value.
- ❑ If the melting point is low the product can be re-crystallized from glacial acetic acid.
- ❑ Obtain an IR spectrum of the product and compare it with that of the starting material.
- ❑ Obtain $^{1}$H-NMR and $^{13}$C-NMR spectra in $d_6$-DMSO if instructed to do so.

## QUESTIONS

1. The classical Perkin reaction involves the reaction of benzaldehyde with acetic anhydride to form cinnamic acid. Write the reaction mechanism for this condensation reaction.

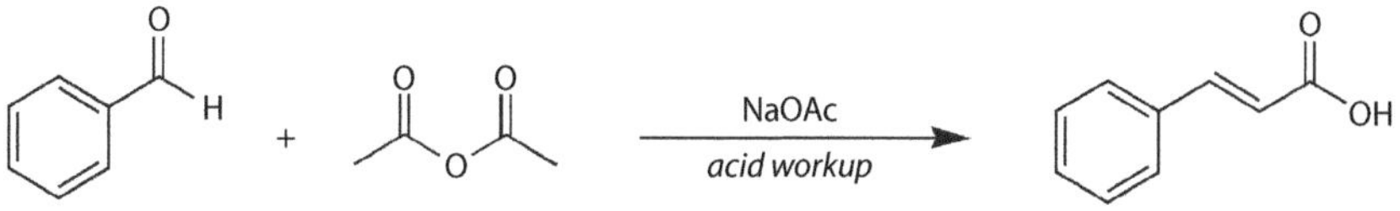

NaOAc
acid workup

*benzaldehyde* *acetic anhydride* *NaOAc = sodium acetate* *cinnamic acid*

2. A carbanion is formed in base-catalyzed aldol condensations and related reactions. Which of the compounds below will be deprotonated by a base the fastest?

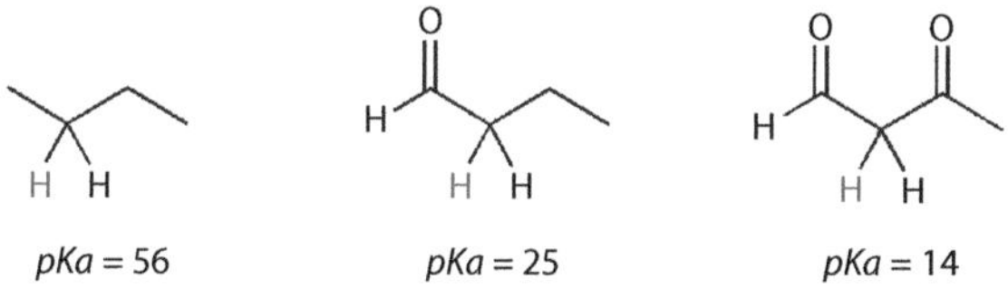

*pKa* = 56 *pKa* = 25 *pKa* = 14

3. Draw the two possible enolate anions for each of the following compounds. Then highlight which is the more stable of the two.

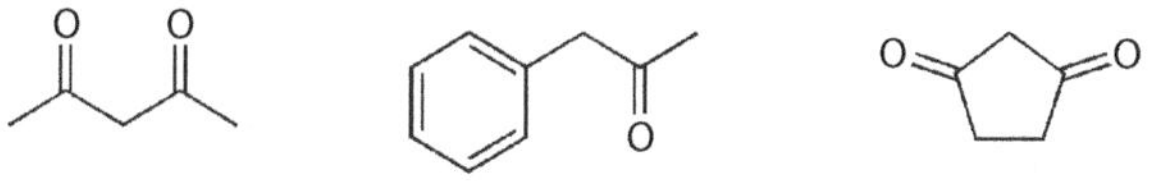

# 8 Williamson Ether Synthesis

## *Preparation of Allyl Phenyl Ether*

**LEARNING GOALS**

- To synthesize an ether compound
- To perform a nucleophilic substitution reaction on a primary allylic halide
- To isolate the product from the reactants using extraction

## INTRODUCTION

There are several methods in the preparation of ethers starting with an alcohol, including

Williamson Ether Synthesis

$$\text{C–OH} \underset{}{\overset{base}{\rightleftharpoons}} \text{C–O}^{\ominus} \xrightarrow{\text{R-X}} \text{C–OR}$$

Acid-Catalyzed Condensation

$$2\ \text{C–OH} \xrightarrow{acid} \text{C–O–C} + H_2O$$

Alkoxymercuration-Demercuration

$$\text{C=C} \xrightarrow[\text{NaBH}_4]{\text{Hg(OAc)}_2\text{/ROH}} \text{H–C–C–OR}$$

The Williamson ether synthesis is considered an easy and fast way to make unsymmetrical ethers. The reaction was first discovered in 1850 when methyl ethyl ether was prepared by reaction of potassium ethoxide with methyl iodide.

The general approach to ether synthesis is to form an alkoxide by reaction of an alcohol with sodium or potassium metal or with sodium hydride. The alkoxide is then added to a primary or unhindered halo-organic compound. The oxygen anion displaces the halogen in an $S_N2$-type reaction.

This experiment uses phenol, an aromatic alcohol, as a substrate instead of a saturated alcohol. Phenol is slightly acidic, having a pKa = 9.95 compared to the pKa of saturated alcohols of 16–20. Potassium carbonate can be used as the base to form the alkoxide instead of the stronger bases traditionally used. The phenoxide anion is a strong electron donor, due to the lone pairs of electrons on the oxygen atom, and thus acts as a nucleophile.

$$C_6H_5OH + K_2CO_3 \rightleftharpoons C_6H_5O^{-}K^{+} + KCO_3H$$

The alkyl halide used is 1-bromoprop-2-ene (allyl bromide), a primary allylic halide. The allylic carbon-1 is electrophilic because it is bonded to the electronegative bromine atom. Electron density is drawn away, giving the carbon a partially positive charge. The conjugation also gives a partial-positive charge to carbon-3. The phenoxide anion is attracted to the partially positive carbons and can react either at carbon-1 in a direct nucleophilic substitution or at carbon-1 eliminating the bromide ion through an allylic rearrangement. The same product, allyl phenyl ether ([2-prepenyloxy]benzene), is produced by both mechanisms.

$$C_6H_5O^{-} + CH_2BrCH{=}CH_2 \longrightarrow C_6H_5OCH_2CH{=}CH_2 + Br^{-}$$

Reaction at carbon-1.

$$C_6H_5O^{-} + CH_2{=}CHCH_2Br \longrightarrow C_6H_5OCH_2CH{=}CH_2 + Br^{-}$$

Reaction at carbon-3.

## PROCEDURE FOR USE IN A MONOMODE MICROWAVE UNIT

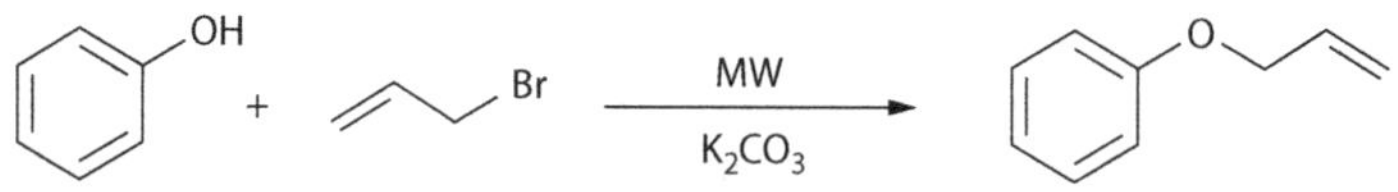

**Caution:** Gloves should be worn while measuring out the reagents. Phenol and allyl bromide are lachrymatory and irritants. They should be measured out in a fume cupboard or well-ventilated area. The use of goggles with side-shields, lab coats, and gloves is considered minimum and nondiscretionary safety practice in the laboratory.

### Table of Reagents and Physical Constants

| Reagent | Equiv. | FW | mmol | Mass (mg) | Density (g/mL) | Vol. (mL) | mp/bp°C |
|---|---|---|---|---|---|---|---|
| Phenol $C_6H_6O$ | 1 | 94.0 | 1.13 | 106 | — | — | 40–42 |
| Allyl bromide $C_3H_5Br$ | 1 | 123 | 1.13 | 136 | 1.398 | 0.097 | 71 |
| Potassium carbonate $K_2CO_3$ | 2 | 138 | 2.26 | 312 | — | — | — |

### Perform the Reaction

- ❑ In a 10-mL glass microwave reaction vessel place a magnetic stir bar and the phenol.
- ❑ Using an automatic delivery pipette add the allyl bromide to the reaction vessel.
- ❑ Add the potassium carbonate to the reaction vessel.
- ❑ Seal the reaction vessel with a cap according to the microwave manufacturer's recommendations.
- ❑ Place the sealed reaction vessel in the microwave cavity.
- ❑ Program the microwave unit to heat the vessel contents to 120°C over a 2-minute ramp period and then hold at this temperature for 20 minutes.
- ❑ After the heating step is completed, allow the contents of the reaction vessel to cool to 50°C or below before removing it from the microwave cavity.

### Isolate the Product

- ❑ While the reaction is cooling, obtain four 25-mL Erlenmeyer flasks and label them "aqueous phase", "wash phase", "organic phase 1", and "organic phase 2".

- ❑ Prepare a TLC plate, approximately 2.5 cm × 7.5 cm in size, for analysis of the product. Draw a light pencil line across the short side of the plate, about 5–10 mm from the bottom.
- ❑ Carefully open the cooled reaction vessel.
- ❑ Add water (5 mL) to the contents of the reaction vessel to dissolve any solids.
- ❑ Transfer the solution to a 30-mL or 60-mL separatory funnel.
- ❑ Rinse the reaction vessel with diethyl ether (3 mL).
- ❑ Transfer the diethyl ether washings to the separatory funnel.
- ❑ Carefully stopper the separatory funnel and invert the funnel.
- ❑ Immediately vent the funnel by opening the stopcock to release pressure that may have developed.
- ❑ Close the stopcock and mix the two layers several times by inverting the funnel repeatedly.
- ❑ Vent the funnel.
- ❑ Close the stopcock and re-clamp the funnel to a ring stand and remove the stopper.
- ❑ Allow the layers to separate.
- ❑ Move the bottom aqueous layer through the stopcock into the 25-mL Erlenmeyer flask labeled "aqueous phase".
- ❑ Pour the ether layer through the top of the separatory funnel into the 25-mL Erlenmeyer flask labeled "organic phase 1".
- ❑ Return the aqueous layer to the separatory funnel.
- ❑ Extract the aqueous layer by adding diethyl ether (3 mL).
- ❑ Carefully stopper the separatory funnel and invert the funnel.
- ❑ Immediately vent the funnel by opening the stopcock to reduce pressure that may have developed in the funnel.
- ❑ Close the stopcock and mix the two layers several times by inverting the funnel repeatedly.
- ❑ Vent the funnel.
- ❑ Close the stopcock and re-clamp the funnel to a ring stand and remove the stopper.
- ❑ Allow the layers to separate.
- ❑ Move the bottom aqueous layer through the stopcock into the 25-mL Erlenmeyer flask labeled "aqueous phase".
- ❑ Pour the ether layer through the top of the separatory funnel into the 25-mL Erlenmeyer flask labeled "organic phase 1" that already contains the previous ether layer.
- ❑ Return the aqueous layer to the separatory funnel.
- ❑ Repeat the extraction of the aqueous layer with 3-mL diethyl ether collecting the ether layer in the Erlenmeyer flask labeled "organic phase 1".
- ❑ Rinse the separatory funnel with water and acetone discarding the rinses in appropriate waste containers.
- ❑ Clamp the separatory funnel to the ring stand and close the stopcock.
- ❑ Return to the separatory funnel the combined ether layers from the Erlenmeyer flask labeled "organic phase 1" (approximately 9 mL).

- ❑ Add 2 M sodium hydroxide solution (5 mL) to the separatory funnel to remove any unreacted phenol.
- ❑ Carefully stopper the separatory funnel and invert the funnel.
- ❑ Immediately vent the funnel by opening the stopcock to reduce pressure that may have developed in the funnel.
- ❑ Close the stopcock and mix the two layers several times by inverting the funnel repeatedly.
- ❑ Vent the funnel.
- ❑ Close the stopcock and re-clamp the funnel to a ring stand and remove the stopper.
- ❑ Allow the layers to separate.
- ❑ Move the bottom aqueous layer through the stopcock into the 25-mL Erlenmeyer flask labeled "wash phase".
- ❑ Add distilled water (5 mL) to the separatory funnel to "wash" the organic layer.
- ❑ Carefully stopper the separatory funnel and invert the funnel.
- ❑ Immediately vent the funnel by opening the stopcock to reduce pressure that may have developed in the funnel.
- ❑ Close the stopcock and mix the two layers several times by inverting the funnel repeatedly.
- ❑ Vent the funnel.
- ❑ Close the stopcock and re-clamp the funnel to a ring stand and remove the stopper.
- ❑ Allow the layers to separate.
- ❑ Move the bottom aqueous layer through the stopcock into the 25-mL Erlenmeyer flask labeled "wash phase".
- ❑ Repeat the wash steps with saturated sodium chloride solution (5 mL).
- ❑ Pour the organic phase into the clean 25-mL Erlenmeyer flask labeled "organic phase 2".
- ❑ Dry this organic layer by adding to the flask containing the ether layer, a drying reagent (500 mg), either anhydrous magnesium sulfate ($MgSO_4$) or anhydrous sodium sulfate ($Na_2SO_4$).
- ❑ Decant the diethyl ether solution into a tared 25-mL or 50-mL round-bottom flask leaving the drying agent behind.
- ❑ Rinse the Erlenmeyer flask and drying reagent by adding to the flask diethyl ether (3 mL).
- ❑ Swirl the flask.
- ❑ Decant the ether rinse into the round bottom flask.
- ❑ Using a micropipette or TLC spotter remove a small aliquot from the round-bottom flask and spot the prepared TLC plate.
- ❑ Remove the ether solvent under reduced pressure until a constant weight is reached.
- ❑ Re-weigh the flask containing the oil.
- ❑ Calculate the yield and percent yield.

## Characterize the Product

- ❑ To determine the purity of the product, run a TLC of the sample using 20% ethyl acetate in hexane as the eluent.
- ❑ Obtain an IR spectrum of the product and compare it with that of the starting material.
- ❑ Obtain $^1$H-NMR and $^{13}$C-NMR spectra in $CDCl_3$ if instructed to do so.

## PROCEDURE FOR USE IN A MULTIMODE MICROWAVE UNIT

Phenol + allyl bromide →(MW, $K_2CO_3$) allyl phenyl ether

**Caution:** Gloves should be worn while measuring out the reagents. Phenol and allyl bromide are lachrymatory and irritants. They should be measured out in a fume cupboard or well-ventilated area. The use of goggles with side-shields, lab coats, and gloves is considered minimum and nondiscretionary safety practice in the laboratory.

### Table of Reagents and Physical Constants

| Reagent | Equiv. | FW | mmol | Mass (mg) | Density (g/mL) | Vol. (mL) | mp/bp°C |
|---|---|---|---|---|---|---|---|
| Phenol $C_6H_6O$ | 1 | 94.0 | 3.08 | 290 | — | — | 40–42 |
| Allyl bromide $C_3H_5Br$ | 1 | 121 | 3.08 | 373 | 1.398 | 0.267 | 71 |
| Potassium carbonate $K_2CO_3$ | 2 | 138 | 6.16 | 850 | — | — | — |

### Perform the Reaction

- ❑ In a 25-mL glass microwave reaction vessel place a magnetic stir bar and the phenol.
- ❑ Using an automatic delivery pipette add the allyl bromide to the reaction vessel.
- ❑ Add the potassium carbonate to the reaction vessel.
- ❑ Seal the reaction vessel with a cap according to the microwave manufacturer's recommendations.
- ❑ Place the sealed reaction vessel in the carousel, noting the vessel's position number and ensuring that vessels are evenly spaced around the carousel.
- ❑ When all the group's reaction vessels are in place, load the carousel into the microwave cavity.
- ❑ If provided by the manufacturer, connect the temperature probe to the control vessel.
- ❑ Program the microwave unit to heat the vessel contents to 120°C over a 2-minute ramp period and then hold at this temperature for 20 minutes.
- ❑ After the heating step is completed, allow the contents of the reaction vessel to cool to 50°C or below before removing it from the microwave cavity.

## Isolate the Product

- ❑ While the reaction is cooling, obtain four 50-mL Erlenmeyer flasks and label them "aqueous phase", "wash phase", "organic phase 1", and "organic phase 2".
- ❑ Prepare a TLC plate, approximately 2.5 cm × 7.5 cm in size, for analysis of the product. Draw a light pencil line across the short side of the plate, about 5–10 mm from the bottom.
- ❑ Carefully open the cooled reaction vessel.
- ❑ Add water (10 mL) to the contents of the reaction vessel to dissolve any solids.
- ❑ Transfer the solution to a 60-mL separatory funnel.
- ❑ Rinse the reaction vessel with diethyl ether (5 mL).
- ❑ Transfer the diethyl ether to the separatory funnel.
- ❑ Carefully stopper the separatory funnel and invert the funnel.
- ❑ Immediately vent the funnel by opening the stopcock to release pressure that may have developed.
- ❑ Close the stopcock and mix the two layers several times by inverting the funnel repeatedly.
- ❑ Vent the funnel.
- ❑ Close the stopcock and re-clamp the funnel to a ring stand and remove the stopper.
- ❑ Allow the layers to separate.
- ❑ Move the bottom aqueous layer through the stopcock into the 50-mL Erlenmeyer flask labeled "aqueous phase".
- ❑ Pour the ether layer through the top of the separatory funnel into a 50-mL Erlenmeyer flask labeled "organic phase 1".
- ❑ Return the aqueous layer to the separatory funnel.
- ❑ Extract the aqueous layer by adding diethyl ether (5 mL).
- ❑ Carefully stopper the separatory funnel and invert the funnel.
- ❑ Immediately vent the funnel by opening the stopcock to reduce pressure that may have developed in the funnel.
- ❑ Close the stopcock and mix the two layers several times by inverting the funnel repeatedly.
- ❑ Vent the funnel.
- ❑ Close the stopcock and re-clamp the funnel to a ring stand and remove the stopper.
- ❑ Allow the layers to separate.
- ❑ Move the bottom aqueous layer through the stopcock into the 50-mL Erlenmeyer flask labeled "aqueous phase".
- ❑ Pour the ether layer through the top of the separatory funnel into the 50-mL Erlenmeyer flask labeled "organic phase 1" that already contains the previous ether layer.
- ❑ Return the aqueous layer to the separatory funnel.

- ❑ Repeat the extraction of the aqueous layer with 5-mL diethyl ether collecting the ether layer in the Erlenmeyer labeled "organic phase 1".
- ❑ Rinse the separatory funnel with water and acetone, discarding the rinses in appropriate waste containers.
- ❑ Clamp the separatory funnel to the ring stand and close the stopcock.
- ❑ Return to the separatory funnel the combined ether layers from the Erlenmeyer flask labeled "organic phase 1" (approximately 15 mL).
- ❑ Add 2 M sodium hydroxide solution (5 mL) to the separatory funnel to remove any unreacted phenol.
- ❑ Carefully stopper the separatory funnel and invert the funnel.
- ❑ Immediately vent the funnel by opening the stopcock to reduce pressure that may have developed in the funnel.
- ❑ Close the stopcock and mix the two layers several times by inverting the funnel repeatedly.
- ❑ Vent the funnel.
- ❑ Close the stopcock and re-clamp the funnel to a ring stand and remove the stopper.
- ❑ Allow the layers to separate.
- ❑ Remove the bottom aqueous layer through the stopcock into the 50-mL Erlenmeyer flask labeled "wash phase".
- ❑ Add distilled water (5 mL) to the separatory funnel to "wash" the organic layer.
- ❑ Carefully stopper the separatory funnel and invert the funnel.
- ❑ Immediately vent the funnel by opening the stopcock to reduce pressure that may have developed in the funnel.
- ❑ Close the stopcock and mix the two layers several times by inverting the funnel repeatedly.
- ❑ Vent the funnel.
- ❑ Close the stopcock and re-clamp the funnel to a ring stand and remove the stopper.
- ❑ Allow the layers to separate.
- ❑ Move the bottom aqueous layer through the stopcock into the 50-mL Erlenmeyer flask labeled "wash phase".
- ❑ Repeat the wash steps with saturated sodium chloride solution (5 mL).
- ❑ Pour the organic phase into the clean 50-mL Erlenmeyer flask labeled "organic phase 2".
- ❑ Dry this organic layer by adding to the flask containing the ether layer, a drying reagent (500 mg), either anhydrous magnesium sulfate ($MgSO_4$) or anhydrous sodium sulfate ($Na_2SO_4$).
- ❑ Decant the diethyl ether solution into a tared 50-mL round bottom flask leaving the drying agent behind.
- ❑ Rinse the Erlenmeyer flask and drying reagent by adding to the flask diethyl ether (5 mL).
- ❑ Swirl the flask.
- ❑ Decant the ether rinse into the round-bottom flask.

- ❑ Using a micropipette or TLC spotter remove a small aliquot from the round-bottom flask and spot the prepared TLC plate.
- ❑ Remove the ether solvent under reduced pressure until a constant weight is reached.
- ❑ Re-weigh the flask containing the oil.
- ❑ Calculate the yield and percent yield.

## Characterize the Product

- ❑ To determine the purity of the product, run a TLC of the sample using 20% ethyl acetate in hexane as the eluent.
- ❑ Obtain an IR spectrum of the product and compare it with that of the starting material.
- ❑ Obtain $^{1}H$-NMR and $^{13}C$-NMR spectra in $CDCl_3$ if instructed to do so.

## QUESTIONS

1. What is the purpose of adding potassium carbonate ($K_2CO_3$) to the reaction?
2. In the preparation of ethyl neopentyl ether there are two possible synthetic routes that involve an alkoxide and a primary halide, shown below. Which one of these two pathways proceeds more rapidly and why?

$$(CH_3)_3C-CH_2-Br + {}^{\ominus}O-CH_2CH_3 \longrightarrow (CH_3)_3C-CH_2-O-CH_2CH_3$$

$$(CH_3)_3C-CH_2-O^{\ominus} + Br-CH_2CH_3 \longrightarrow (CH_3)_3C-CH_2-O-CH_2CH_3$$

3. Determine a structure for the intermediate and product of the following reaction.

$$H_3C-CH=CH_2 + Br_2 + H_2O \longrightarrow \underset{(C_3H_7BrO)}{A} + HBr \xrightarrow{Na} \underset{(C_3H_6O)}{B} + NaBr + 1/2\,H_2$$

4. Write the two possible Williamson ether syntheses for the following ethers and determine if one has an advantage over the other route.
   a. Methyl isopropyl ether
   b. Ethyl isobutyl ether
   c. Hexyl neopentyl ether

# 9 Claisen Rearrangement

## *Preparation of 2-Allyl Phenol from Allyl Phenyl Ether*

**LEARNING GOALS**

- To perform a rearrangement reaction
- To perform a reaction under solvent-free conditions

## INTRODUCTION

Rearrangement reactions are a broad class of organic reactions where the carbon skeleton of a molecule is rearranged to give a structural isomer of the original molecule. An example is the Claisen rearrangement, the first reported [3,3]-sigmatropic rearrangement, discovered in 1912 by Rainer L. Claisen. The net result of a sigmatropic reaction is that one σ-bond is converted to another σ-bond in an intramolecular process.

The conventional method for determining the type of sigmatropic rearrangement is to number the atom of the bond being broken as "1", and then count the atoms in each direction from the broken bond to the atoms that form the new σ-bond in the product, numbering consecutively. The numbers that correspond to the atoms forming the new bond are then separated by a comma and placed within brackets. When allyl vinyl ether is heated, the allyl group moves from being attached to the oxygen to being attached to the terminal carbon of the vinyl group.

bond broken

[3, 3] *sigmatropic rearrangement*

bond formed

*allyl vinyl ether*

The aromatic Claisen rearrangement is a [3 + 3]-sigmatropic rearrangement in which aryl vinyl ether is converted to a vinyl phenol. This reaction forms a carbon–carbon bond between the aromatic carbon and the allylic carbon through a concerted pericyclic reaction. The initial rearrangement forms a ketone that then rapidly tautomerizes to the phenol by re-aromatizing the ring.

*allyl phenyl ether* → [intermediate] → *2-allyl phenol*

## PROCEDURE FOR USE IN A MONOMODE MICROWAVE UNIT

O → MW → OH

**Caution:** The starting material and the product (allylphenyl ether and 2-allylphenol) are corrosive and should be handled with care. The use of goggles with side-shields, lab coats, and gloves is considered minimum and nondiscretionary safety practice in the laboratory.

### Table of Reagents and Physical Constants

| Reagent | Equiv. | FW | mmol | Mass (mg) | Density (g/mL) | Vol. (mL) | mp/bp°C |
|---|---|---|---|---|---|---|---|
| Allyl phenyl ether $C_9H_{10}O$ | 1 | 134 | 3.82 | 512 | 0.978 | 0.523 | 192 |
| Tetrabutylammonium bromide $C_{16}H_{36}BrN$ | 0.13 | 322 | 0.50 | 160 | — | — | 102–106 |

### Perform the Reaction

- ❑ In a 10-mL glass microwave reaction vessel place a magnetic stir bar, the allyl phenyl ether, and the tetrabutylammonium bromide.
- ❑ Seal the reaction vessel with a cap according to the microwave manufacturer's recommendations.
- ❑ Place the sealed reaction vessel in the microwave cavity.
- ❑ Program the microwave unit to heat the vessel contents to 245°C using full power and then hold at this temperature for 30 minutes.
- ❑ After the heating step is completed, allow the contents of the reaction vessel to cool to 50°C or below before removing it from the microwave cavity.

### Isolate the Product

- ❑ During the reaction time obtain a separatory funnel (60 or 125 mL) and clamp it on a ring stand. Also obtain three 25-mL Erlenmeyer flasks and label them "aqueous phase", "organic phase 1", and "organic phase 2".
- ❑ Carefully open the cooled reaction vessel.
- ❑ Add 2N NaOH (5 mL) to the reaction vessel.
- ❑ Transfer the contents of the microwave vessel to the separatory funnel (check that the stopcock is closed).
- ❑ Add diethyl ether (3 mL) to the separatory funnel to extract the unreacted starting material.

- ❑ Carefully stopper the separatory funnel and invert the funnel.
- ❑ Immediately vent the funnel by opening the stopcock to release pressure that may have developed.
- ❑ Close the stopcock and mix the two layers several times by inverting the funnel repeatedly.
- ❑ Vent the funnel.
- ❑ Close the stopcock and re-clamp the funnel to a ring stand and remove the stopper.
- ❑ Allow the layers to separate.
- ❑ Move the bottom aqueous layer through the stopcock into the 25-mL Erlenmeyer flask labeled "aqueous phase".
- ❑ Pour the ether layer through the top of the separatory funnel into the 25-mL Erlenmeyer flask labeled "organic phase 1".
- ❑ Return the basic aqueous layer to the separatory funnel.
- ❑ Repeat the extraction with diethyl ether (3 mL).
- ❑ After the layers have separated transfer the lower aqueous phase to the Erlenmeyer flask labeled "aqueous phase".
- ❑ Pour the ether layer into the Erlenmeyer flask labeled "organic phase 1" combining it with the first ether layer.
- ❑ Acidify the aqueous layer in the Erlenmeyer flask by adding 6 M hydrochloric acid dropwise until the solution has a pH of 4–5 as determined on broad-range pH paper.
- ❑ Transfer the aqueous phase back into the separatory funnel.
- ❑ Add diethyl ether (5 mL) to the separatory funnel to extract the product from the aqueous phase.
- ❑ Carefully stopper the separatory funnel and invert the funnel.
- ❑ Immediately vent the funnel by opening the stopcock to release pressure that may have developed.
- ❑ Close the stopcock and mix the two layers several times by inverting the funnel repeatedly.
- ❑ Vent the funnel.
- ❑ Close the stopcock and re-clamp the funnel to a ring stand and remove the stopper.
- ❑ Allow the layers to separate.
- ❑ Move the bottom aqueous layer through the stopcock into the 25-mL Erlenmeyer flask labeled "aqueous phase".
- ❑ Pour the ether layer through the top of the separatory funnel into the clean 25-mL Erlenmeyer flask labeled "organic phase 2".
- ❑ Return the aqueous layer to the separatory funnel.
- ❑ Repeat the extraction of the aqueous layer with diethyl ether (3 mL) twice, collecting the ether layers each time in the Erlenmeyer flask labeled "organic phase 2".
- ❑ Pour the combined organic phase 2 contents back into the separatory funnel.
- ❑ Wash the organics by adding saturated aqueous sodium chloride (NaCl, 5 mL) to the separatory funnel.

- ❑ Carefully stopper the separatory funnel and invert the funnel.
- ❑ Immediately vent the funnel by opening the stopcock to release pressure that may have developed.
- ❑ Close the stopcock and mix the two layers several times by inverting the funnel repeatedly.
- ❑ Vent the funnel.
- ❑ Close the stopcock and re-clamp the funnel to a ring stand and remove the stopper.
- ❑ Allow the layers to separate.
- ❑ Move the bottom aqueous layer through the stopcock into the 25-mL Erlenmeyer flask labeled "aqueous phase".
- ❑ Pour the organic phase into a clean 25-mL Erlenmeyer flask.
- ❑ Dry the organic phase by adding a drying reagent (500 mg), either anhydrous magnesium sulfate ($MgSO_4$) or anhydrous sodium sulfate ($Na_2SO_4$).
- ❑ Decant the dried organics into a tared 25-mL or 50-mL round-bottom flask, being careful to leave the drying agent behind.
- ❑ Rinse the Erlenmeyer flask and drying reagent by adding diethyl ether (3 mL) to the flask.
- ❑ Swirl the flask.
- ❑ Decant the rinse into the round-bottom flask.
- ❑ To determine the purity of the product, run a TLC of this solution by removing a small sample from the round-bottom flask with a micropipette.
- ❑ Spot the solution on a prepared TLC plate.
- ❑ Prepare a TLC chamber by taking a beaker large enough to fit the TLC plate and adding a small layer of 20% ethyl acetate in hexane.
- ❑ Cover the beaker with a watch glass.
- ❑ Carefully place the TLC plate in the beaker leaning the plate against the side of the beaker.
- ❑ Be sure the spot is above the level of the eluting solvent.
- ❑ Re-cover the beaker.
- ❑ Once the solvent front is near the top of the plate carefully remove the plate from the chamber.
- ❑ Immediately mark the solvent front with a pencil and then allow the plate to dry.
- ❑ Visualize the TLC plate with a UV lamp and mark any spots observed.
- ❑ Remove the diethyl ether solvent under reduced pressure until a constant weight is reached.
- ❑ Re-weigh the flask containing the oil.
- ❑ Calculate the yield and percent yield.

## Characterize the Product

- ❑ Obtain an IR spectrum of the product and compare it with that of the starting material.
- ❑ Obtain $^{1}H$-NMR and $^{13}C$-NMR spectra in $CDCl_3$ if instructed to do so.

## PROCEDURE FOR USE IN A MULTIMODE MICROWAVE UNIT

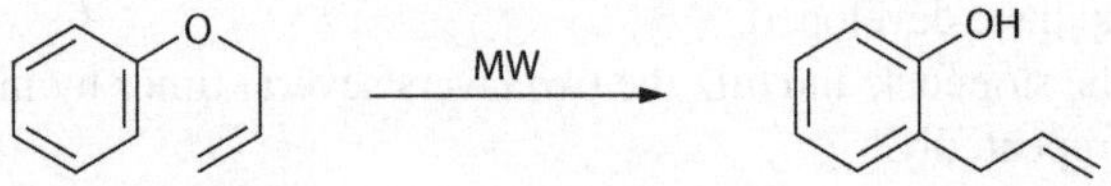

**Caution:** The starting material and the product (allylphenyl either and 2-allylphenol) are corrosive and should be handled with care. The use of goggles with side-shields, lab coats, and gloves is considered minimum and nondiscretionary safety practice in the laboratory.

### Table of Reagents and Physical Constants

| Reagent | Equiv. | FW | mmol | Mass (mg) | Density (g/mL) | Vol. (mL) | mp/bp°C |
|---|---|---|---|---|---|---|---|
| Allyl phenyl ether $C_9H_{10}O$ | 1 | 134 | 7.30 | 978 | 0.978 | 1.0 | 192 |
| Tetrabutylammonium bromide $C_{16}H_{36}BrN$ | 0.13 | 322 | 1.0 | 322 | — | — | 102–106 |

### Perform the Reaction

- ❑ In a 25-mL glass microwave reaction vessel place a magnetic stir bar, the allyl phenyl ether, and the tetrabutylammonium bromide.
- ❑ Seal the reaction vessel with a cap according to the microwave manufacturer's recommendations.
- ❑ Place the sealed reaction vessel in the carousel, noting the vessel's position number and ensuring that vessels are evenly spaced around the carousel.
- ❑ When all the group's reaction vessels are in place, load the carousel into the microwave cavity.
- ❑ If provided by the manufacturer, connect the temperature probe to the control vessel.
- ❑ Program the microwave unit to heat the vessel contents to 245°C with full power and then hold at this temperature for 30 minutes.
- ❑ After the heating step is completed, allow the contents of the reaction vessel to cool to 50°C or below before removing it from the microwave cavity.

### Isolate the Product

- ❑ During the reaction time obtain a separatory funnel (60 or 125 mL) and clamp it on a ring stand. Also obtain three 50-mL Erlenmeyer flasks and label them "aqueous phase", "organic phase 1", and "organic phase 2".
- ❑ Carefully open the cooled reaction vessel.

- ❑ Add 2N NaOH (10 mL) to the reaction vessel.
- ❑ Transfer the contents of the microwave vessel to the separatory funnel (check that the stopcock is closed).
- ❑ Add diethyl ether (5 mL) to the separatory funnel to extract the unreacted starting material.
- ❑ Carefully stopper the separatory funnel and invert the funnel.
- ❑ Immediately vent the funnel by opening the stopcock to release pressure that may have developed.
- ❑ Close the stopcock and mix the two layers several times by inverting the funnel repeatedly.
- ❑ Vent the funnel.
- ❑ Close the stopcock and re-clamp the funnel to a ring stand and remove the stopper.
- ❑ Allow the layers to separate.
- ❑ Move the bottom aqueous layer through the stopcock into the 50-mL Erlenmeyer flask labeled "aqueous phase".
- ❑ Pour the ether layer through the top of the separatory funnel into the 50-mL Erlenmeyer flask labeled "organic phase 1".
- ❑ Return the basic aqueous layer to the separatory funnel.
- ❑ Repeat the extraction with diethyl ether (5 mL).
- ❑ After the layers have separated transfer the lower aqueous phase to the Erlenmeyer flask labeled "aqueous phase".
- ❑ Pour the ether layer into the Erlenmeyer flask labeled "organic phase 1" combining it with the first ether layer.
- ❑ Acidify the aqueous layer in the Erlenmeyer flask by adding 6 M hydrochloric acid dropwise until the solution has a pH of 4–5 as determined on broad-range pH paper.
- ❑ Transfer the aqueous phase back into the separatory funnel.
- ❑ Add diethyl ether (5 mL) to the separatory funnel to extract the product from the aqueous phase.
- ❑ Carefully stopper the separatory funnel and invert the funnel.
- ❑ Immediately vent the funnel by opening the stopcock to release pressure that may have developed.
- ❑ Close the stopcock and mix the two layers several times by inverting the funnel repeatedly.
- ❑ Vent the funnel.
- ❑ Close the stopcock and re-clamp the funnel to a ring stand and remove the stopper.
- ❑ Allow the layers to separate.
- ❑ Move the bottom aqueous layer through the stopcock into the 50-mL Erlenmeyer flask labeled "aqueous phase".
- ❑ Pour the ether layer through the top of the separatory funnel into the clean 50-mL Erlenmeyer flask labeled "organic phase 2".
- ❑ Return the aqueous layer to the separatory funnel.

- ❑ Repeat the extraction of the aqueous layer with diethyl ether (3 mL) twice, collecting the ether layers each time in the Erlenmeyer flask labeled "organic phase 2".
- ❑ Pour the combined organic phase 2 contents back into the separatory funnel.
- ❑ Wash the organics by adding saturated aqueous sodium chloride (NaCl, 5 mL) to the separatory funnel.
- ❑ Carefully stopper the separatory funnel and invert the funnel.
- ❑ Immediately vent the funnel by opening the stopcock to release pressure that may have developed.
- ❑ Close the stopcock and mix the two layers several times by inverting the funnel repeatedly.
- ❑ Vent the funnel.
- ❑ Close the stopcock and re-clamp the funnel to a ring stand and remove the stopper.
- ❑ Allow the layers to separate.
- ❑ Move the bottom aqueous layer through the stopcock into the 25-mL Erlenmeyer flask labeled "aqueous phase".
- ❑ Pour the organic phase into a clean 50-mL Erlenmeyer flask.
- ❑ Dry the organic phase by adding a drying reagent (500 mg), either anhydrous magnesium sulfate ($MgSO_4$) or anhydrous sodium sulfate ($Na_2SO_4$).
- ❑ Decant the dried organics into a tared 50-mL round-bottom flask, being careful to leave the drying agent behind.
- ❑ Rinse the Erlenmeyer flask and drying reagent by adding diethyl ether (3 mL) to the flask.
- ❑ Swirl the flask.
- ❑ Decant the rinse into the round-bottom flask.
- ❑ To determine the purity of the product, run a TLC of this solution by removing a small sample from the round-bottom flask with a micropipette.
- ❑ Spot the solution on a prepared TLC plate.
- ❑ Prepare a TLC chamber by taking a beaker large enough to fit the TLC plate and adding a small layer of 20% ethyl acetate in hexane.
- ❑ Cover the beaker with a watch glass.
- ❑ Carefully place the TLC plate in the beaker leaning the plate against the side of the beaker.
- ❑ Be sure the spot is above the level of the eluting solvent.
- ❑ Re-cover the beaker.
- ❑ Once the solvent front is near the top of the plate carefully remove the plate from the chamber.
- ❑ Immediately mark the solvent front with a pencil and then allow the plate to dry.
- ❑ Visualize the TLC plate with a UV lamp and mark any spots observed.
- ❑ Remove the diethyl ether solvent under reduced pressure until a constant weight is reached.
- ❑ Re-weigh the flask containing the oil.
- ❑ Calculate the yield and percent yield.

## Characterize the Product

- ❑ Obtain an IR spectrum of the product and compare it with that of the starting material.
- ❑ Obtain $^{1}$H-NMR and $^{13}$C-NMR spectra in $CDCl_3$ if instructed to do so.

## QUESTIONS

1. Show how the electrons flow in the following [3,3]-sigmatropic rearrangement.

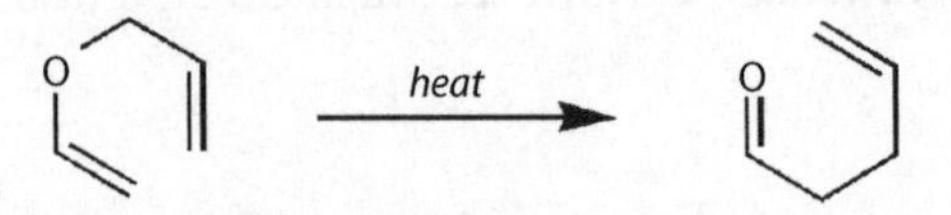


2. Define "concerted pericyclic reaction".
3. Define the following rearrangement (i.e., determine x and y in [x,y]-rearrangement).

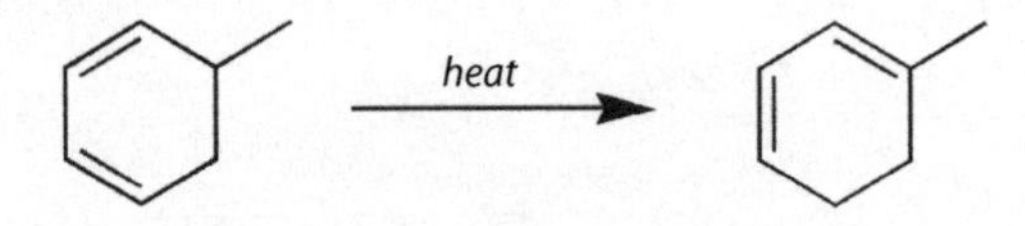


4. Why was 2N NaOH added to the reaction mixture instead of neutral water to remove the unreacted starting material?
5. What was the purpose of washing the organic phase 2 with an aqueous solution of saturated sodium chloride?

# 10 Hydration of an Alkyne

## *Preparation of Acetophenone from Phenylacetylene*[*]

**LEARNING GOALS**

- To synthesize a compound, taking advantage of enol to keto tautomerism
- To use an internal standard and gas chromatography–mass spectroscopy to calculate yield

## INTRODUCTION

The electrophilic addition of water is one of the most common methods for functionalization of alkenes and alkynes. In the case of alkenes, the reaction can be catalyzed by a strong Brønsted acid, such as sulfuric acid, yielding an alcohol whose structure follows Markovnikov regiochemistry. The usefulness of acid-catalyzed alkene hydration is limited to terminal and symmetric starting materials because mixtures of isomers are formed when using nonsymmetric alkenes. Another problem is that rearrangements can take place during the course of the reaction, producing a range of unwanted side-products. This can be avoided by performing the reaction under neutral conditions in the presence of a mercury salt, typically mercury(II) acetate, in what is known as oxymercuration. This is followed by treatment of the product with a reducing agent such as sodium borohydride to remove the mercury group (called demercuration). The addition of water to alkynes also requires a strong acid and is facilitated by mercury salts. However, unlike the additions to double bonds that give alcohol products, addition of water to alkynes gives ketone products. This is the result of the tautomerization of the initial enol product, an equilibrium that lies far toward the direction of the carbonyl tautomer due to its greater stability.

---

[*] Experimental procedure developed by Dr. Javier Horta, Merrimack College. Modified from a conventional procedure: J. P. Damiano and M. Postel, *J. Organomet. Chem.*, 1996, *522*, 303–305.

$H_2O/H^+$, $Hg(OAc)_2$ — *oxymercuration*; Hg(OAc); $NaBH_4$ — *demercuration*; $CH_3$

$H_2O/H^+$, $HgSO_4$; *tautomerization*; *enol form*; *keto form*

The high toxicity of mercury and its compounds has warranted the exploration of alternative ways to perform the hydration of alkenes and alkynes. Iron(III) salts can facilitate the hydration of terminal aromatic alkynes to ketones. One example is to use $FeCl_3{\cdot}6H_2O$, which can serve as both the metal reagent and also the source of water.

$FeCl_3 \cdot 6H_2O$; $CH_2R$

## PROCEDURE FOR USE IN A MONOMODE MICROWAVE UNIT

MW

$FeCl_3 \cdot 6H_2O$

**Caution:** Iron(III) chloride is a corrosive solid. Phenylacetylene and dichloromethane are irritants. The use of goggles with side-shields, lab coats, and gloves is considered minimum and nondiscretionary safety practice in the laboratory.

### Table of Reagents and Physical Constants

| Reagent | Equiv. | FW | mmol | Mass (mg) | Density (g/mL) | Vol. (mL) | mp/bp°C |
|---|---|---|---|---|---|---|---|
| Phenylacetylene $C_8H_6$ | 1 | 102 | 0.775 | 79.1 | 0.930 | 0.085 | 142–144 |
| Iron(III) chloride hexahydrate $FeCl_3 \cdot 6H_2O$ | 1 | 270 | 0.778 | 210 | — | 1.82 | 37 |
| Octane $C_8H_{18}$ | — | — | — | — | — | 0.085 | 125–126 |
| Dichloromethane $CH_2Cl_2$ | — | — | — | — | — | 5 | 40 |

### Perform the Reaction

- ❑ In a 25-mL glass microwave reaction vessel place a magnetic stir bar and the dichloromethane.
- ❑ Using an Eppendorf pipette add the phenylacetylene and the internal standard octane.
- ❑ Stir the solution gently on a magnetic stir plate to mix the reagents.
- ❑ Insert the tip of a short Pasteur pipette into the reaction mixture to collect a very small aliquot and transfer it to a small test tube.
- ❑ Wash the pipette into the small test tube with dichloromethane (1 mL); this sample will serve as the control for the gas chromatography–mass spectrometry (GC/MS).
- ❑ Add the iron(III) chloride hexahydrate to the reaction vessel.
- ❑ Seal the reaction vessel with a cap according to the microwave manufacturer's recommendations.
- ❑ Place the sealed reaction vessel in the microwave cavity.
- ❑ Program the microwave unit to heat the vessel contents to 110°C over a 5-minute period and hold at this temperature for 15 minutes.
- ❑ After the heating step is completed, allow the contents of the reaction vessel to cool to 50°C or below before removing it from the microwave cavity.

## Isolate the Product

- ❑ While the reaction mixture cools, use the tip of a long Pasteur pipette to insert a small piece of tissue paper into the neck of a short Pasteur pipette so as to plug its outlet where it begins to narrow.
- ❑ Add a small amount of silica gel into the plugged Pasteur pipette (about a quarter of the volume of the plugged pipette).
- ❑ Clamp the pipette to a ring stand or lab bars with the tip of the pipette inserted into a small test tube.
- ❑ Carefully open the reaction vessel.
- ❑ As performed with the initial mixture, insert the tip of another short Pasteur pipette into the reaction mixture so as to collect a very small aliquot.
- ❑ Place the pipette containing the aliquot into the top of the plugged pipette containing silica gel.
- ❑ Use a third pipette to add dichloromethane (1 mL) into the pipette containing the aliquot so that the aliquot and dichloromethane pass through the silica gel.*
- ❑ The solution collected in the small test tube under the pipette represents the reaction sample.

## Product Analysis

- ❑ Run the control sample (1 μL) in the GC/MS as instructed, using an initial temperature of 30°C and increasing the temperature by 10°C/min to a final temperature of 130°C.
- ❑ Identify the peaks on the chromatogram corresponding to the substrate and the internal standard by the corresponding mass spectra.
- ❑ Determine the area of the peaks and calculate the ratio of substrate to internal standard.
- ❑ Run the reaction sample in the GC/MS as before.
- ❑ Identify the peaks on the chromatogram corresponding to the product and internal standard by the corresponding mass spectra.
- ❑ Determine the area of the peaks and calculate the ratio of product to internal standard.
- ❑ Determine the percent yield of the reaction by determining the ratio of product to substrate, each relative to the amount of internal standard used.

* If no solution passes through the silica into the small test tube position beneath the pipette, add a further aliquot of dichloromethane (1 mL) to the pipette.

## PROCEDURE FOR USE IN A MULTIMODE MICROWAVE UNIT

MW
$FeCl_3 \cdot 6H_2O$

**Caution:** Iron(III) chloride is a corrosive solid. Phenylacetylene and dichloromethane are irritants. The use of goggles with side-shields, lab coats, and gloves is considered minimum and nondiscretionary safety practice in the laboratory.

### Table of Reagents and Physical Constants

| Reagent | Equiv. | FW | mmol | Mass (mg) | Density (g/mL) | Vol. (mL) | mp/bp°C |
|---|---|---|---|---|---|---|---|
| Phenylacetylene $C_8H_6$ | 1 | 102 | 0.775 | 79.1 | 0.930 | 0.085 | 142–144 |
| Iron(III) chloride hexahydrate $FeCl_3{\cdot}6H_2O$ | 1 | 270 | 0.778 | 210 | — | 1.82 | 37 |
| Octane $C_8H_{18}$ | — | — | — | — | — | — | 125–126 |
| Dichloromethane $CH_2Cl_2$ | — | — | — | — | — | 5 | 40 |

### Perform the Reaction

- ❑ In a 25-mL glass microwave reaction vessel place a magnetic stir bar and the dichloromethane.
- ❑ Using an Eppendorf pipette add the phenylacetylene and the internal standard octane.
- ❑ Stir the solution gently on a magnetic stir plate to mix the reagents.
- ❑ Insert the tip of a short Pasteur pipette into the reaction mixture to collect a very small aliquot and transfer it to a small test tube.
- ❑ Wash the pipette into the small test tube with dichloromethane (1 mL); this sample will serve as the control for the GC/MS.
- ❑ Add the iron(III) chloride hexahydrate to the reaction vessel.
- ❑ Seal the reaction vessel with a cap according to the microwave manufacturer's recommendations.
- ❑ Place the sealed reaction vessel in the carousel, noting the vessel's position number and ensuring that the vessels are evenly spaced around the carousel.
- ❑ When all the group's reaction vessels are in place, load the carousel into the microwave cavity.
- ❑ If provided by the manufacturer, connect the temperature probe to the control vessel.

- ❑ Program the microwave unit to heat the vessel contents to 110°C over a 5-minute ramp period and then hold at this temperature for 15 minutes.
- ❑ After the heating step is completed, allow the contents of the reaction vessel to cool to 50°C or below before removing it from the microwave cavity.

## Isolate the Product

- ❑ While the reaction mixture cools, use the tip of a long Pasteur pipette to insert a small piece of tissue paper into the neck of a short Pasteur pipette so as to plug its outlet where it begins to narrow.
- ❑ Add a small amount of silica gel into the plugged Pasteur pipette (about a quarter of the volume of the plugged pipette).
- ❑ Clamp the pipette to a ring stand or lab bars with the tip of the pipette inserted into a small test tube.
- ❑ Carefully open the reaction vessel.
- ❑ As performed with the initial mixture, insert the tip of another short Pasteur pipette into the reaction mixture so as to collect a very small aliquot.
- ❑ Place the pipette containing the aliquot into the top of the plugged pipette containing silica gel.
- ❑ Use a third pipette to add dichloromethane (1 mL) into the pipette containing the aliquot so that the aliquot and dichloromethane pass through the silica gel.*
- ❑ The solution collected in the small test tube under the pipette represents the reaction sample.

## Product Analysis

- ❑ Run the control sample (1 μL) in the GC/MS as instructed, using an initial temperature of 30°C and increasing the temperature by 10°C/min to a final temperature of 130°C.
- ❑ Identify the peaks on the chromatogram corresponding to the substrate and the internal standard by the corresponding mass spectra.
- ❑ Determine the area of the peaks and calculate the ratio of substrate to internal standard.
- ❑ Run the reaction sample in the GC/MS as before.
- ❑ Identify the peaks on the chromatogram corresponding to the product and internal standard by the corresponding mass spectra.
- ❑ Determine the area of the peaks and calculate the ratio of product to internal standard.
- ❑ Determine the percent yield of the reaction by determining the ratio of product to substrate, each relative to the amount of internal standard used.

---

* If no solution passes through the silica into the small test tube positioned beneath the pipette, add a further aliquot of dichloromethane (1 mL) to the pipette.

## QUESTIONS

1. Looking at the mechanism does the addition reaction follow Markovnikov's Rule?
2. Does the iron(III) chloride act as a Lewis acid or Lewis base?
3. Why must the reaction solution be passed through silica gel before being analyzed by GC/MS but the control solution does not? (Hint: what is the silica gel removing from the reaction mixture?)
4. Give the product expected if hex-1-yne is subjected to hydrolysis under the same conditions.
5. When hex-2-yne is hydrolyzed two products are formed in approximately equal amounts, give the two products and explain how they are formed.
6. Show the two different enols that are in equilibrium with 3-phenyl-2-propanol.

# 11 Oxidation of a Secondary Alcohol

## *Preparation of a Ketone*[*]

**LEARNING GOALS**

- To perform the oxidation of a secondary alcohol to a ketone
- To separate the product from the reactants using extraction and chromatography

## INTRODUCTION

In general chemistry courses we are taught that oxidation involves the loss of electrons, often from metals, to form cations. In organic chemistry, oxidation is explained in terms of covalent bonds formed or lost rather than changes in charge or oxidation number. For a working terminology, when dealing with organic compounds we can classify oxidation as

- The addition of more oxygen bonds to carbon
- The addition of more electronegative atoms bonded to carbon
- The loss of hydrogen atoms bonded to carbon

Primary alcohols can be oxidized to aldehydes. They can also be further oxidized to carboxylic acids. Secondary alcohols are oxidized to ketones whereas tertiary alcohols are resistant to oxidation under most conditions. Looking specifically at the case of a secondary alcohol, oxidation involves the loss of two hydrogen atoms: one from the carbon and one from the adjacent oxygen. When these hydrogens are removed a double bond results.

H O H R R′ —*oxidant* ($-H_2$)→ O R R′

[*] Experimental procedure developed by Michael Troy, Merrimack College. Modified from a conventional procedure: M. Hulce and D. W. Marks, *J. Chem. Ed.* 2001, *78*, 66–67.

There are many metal-containing oxidizing agents, including chromium and manganese reagents. These metals are relatively inexpensive but often require a stoichiometric amount of the heavy metal, which generates a substantial amount of waste. Using a metal catalyst and a nonmetal oxidizing reagent such as hydrogen peroxide is considered a more environmentally friendly approach to this reaction. As an example, sodium tungstate ($Na_2WO_4$) can be used as a catalyst. The reaction proceeds via an initial esterification of the secondary alcohol by tungstic acid (the complex formed when sodium tungstate is placed into solution) to form a charged metal ester complex (compound **1**). This complex, after a hydrogen transfer to make compound **2**, eliminates an equivalent amount of water to form a new metal ester (compound **3**). This is followed by an E2 elimination of the hydrogen bonded to the carbon atom next to the oxygen to form the ketone product and a reduced tungsten intermediate ($HWO_3^-$) in a +4 oxidation state. The tungsten ($W^{4+}$) is then oxidized by the hydrogen peroxide to reform tungstic acid ($W^{6+}$), which can then go around the catalytic cycle again.

Also included in this reaction mixture is a phase transfer catalyst, methyltrioctyl-ammonium hydrogen sulfate, $[CH_3 (C_8H_{17})_3N]^+[HSO_4]^-$. The reaction is run in an aqueous solution of hydrogen peroxide in which the sodium tungstate is soluble whereas most alcohols containing four or more carbons are not. The phase transfer catalyst is soluble in both layers due to the ionic nature of the reagent. Thus the ammonium salt shuttles the water-soluble $WO_4^{2-}$ anion into the organic alcohol phase allowing the reaction to occur. The ammonium reagent then shuttles the reduced catalyst back to the aqueous phase to be re-oxidized by the hydrogen peroxide.

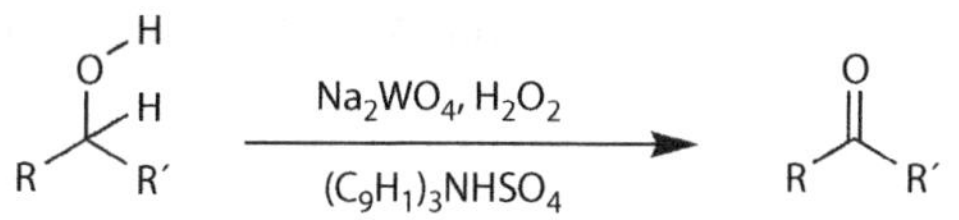

$Na_2WO_4$, $H_2O_2$

$(C_9H_1)_3NHSO_4$

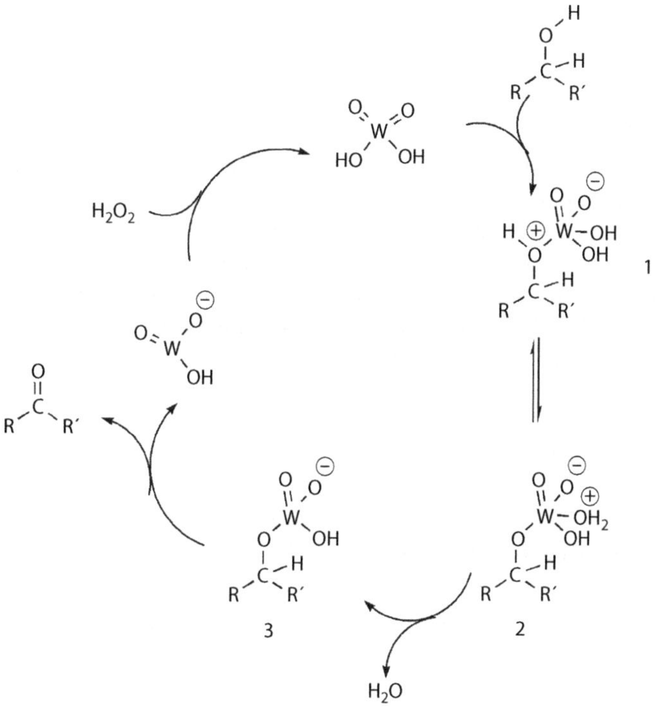

## PROCEDURE FOR USE IN A MONOMODE MICROWAVE UNIT

R(R′)CH–OH $\xrightarrow[(C_9H_1)_3NHSO_4]{Na_2WO_4, H_2O_2}$ R(R′)C=O

Your laboratory instructor will assign you an alcohol to use in this reaction. Prepare your notebook by writing your pre-lab assignment incorporating your assigned alcohol.

**Caution:** Sodium tungstate, methyltrioctylammonium hydrogen sulfate, and 1-phenyl-1-propanol are irritants so avoid skin contact. Aqueous 30% hydrogen peroxide is very corrosive and a strong oxidant. Avoid contact of the reagent with skin and any metal objects such as spatulas. Transfer the hydrogen peroxide solution to a glass graduated cylinder and pipette the required amount directly into the reaction tube. Flush any extra reagent down the sink with large amounts of water. If any peroxide is accidentally spilled on the skin, immediately wash the area with large amounts of water to prevent a burn. The use of goggles with side-shields, lab coats, and gloves is considered minimum and nondiscretionary safety practice in the laboratory.

### Table of Reagents and Physical Constants

| Reagent | Equiv. | FW | mmol | Mass (mg) | Density (g/mL) | Vol. (mL) | mp/bp°C |
|---|---|---|---|---|---|---|---|
| Alcohol ROH | 1 | — | 4.2 | — | — | — | — |
| Sodium tungstate dihydrate $Na_2WO_4{\cdot}H_2O$ | 0.01 | 330 | 0.042 | 14 | — | — | 698 |
| Methyltrioctyl ammonium hydrogen sulfate $C_{67}H_{76}NSO_4$ | 0.01 | 466 | 0.042 | 20 | — | — | 61–65 |
| Hydrogen peroxide (30% aqueous) $H_2O_2$ | — | 34.0 | — | — | — | 0.5 | — |
| ***Alcohols*** | | | | | | | |
| 1-Phenyl-1-propanol $C_9H_{12}O$ | 1 | 136 | 4.2 | 572 | 0.994 | 0.58 | 219 |
| 1-Phenyl-1-ethanol $C_8H_{10}O$ | 1 | 122 | 4.2 | 513 | 1.012 | 0.51 | 19–20 |
| 4-*tert*-Butyl cyclohexanol $C_{10}H_{20}O$ | 1 | 156 | 4.2 | 660 | — | — | 62–70 |

## Perform the Reaction

- ❑ In a 10-mL glass microwave reaction vessel place a magnetic stir bar, the alcohol, sodium tungstate dihydrate, and methyltrioctylammonium hydrogen sulfate.
- ❑ Using an automatic delivery pipette add the 30% hydrogen peroxide to the reaction vessel.
- ❑ Seal the reaction vessel with a cap according to the microwave manufacturer's recommendations.
- ❑ Carefully stir the contents on a magnetic stir plate to dissolve the solid.
- ❑ Place the sealed reaction vessel into the microwave cavity.
- ❑ Program the microwave unit to heat the vessel contents to 150°C over a 2-minute ramp period and then hold at this temperature for 4 minutes.
- ❑ After the heating step is completed, allow the contents of the reaction vessel to cool to 50°C or below before removing it from the microwave cavity.

## Isolate the Product

- ❑ Carefully open the reaction vessel.
- ❑ Add water (1 mL) to the contents of the reaction vessel, followed by diethyl ether (2 mL).
- ❑ Agitate the vessel contents by placing the tube on a magnetic stir plate.
- ❑ Clamp the vessel and allow the contents to separate into two layers.
- ❑ Remove the bottom aqueous layer with a pipette and place the aqueous layer in a 25-mL Erlenmeyer flask.
- ❑ Add a saturated solution of sodium sulfite (2 mL) to the organic layer in the reaction vessel.
- ❑ Agitate the contents of the vessel again.
- ❑ Allow the layers to separate and remove the bottom aqueous layer as before.
- ❑ With a magnetic retrieving wand remove the stir bar.
- ❑ Prepare a "drying pipette" by plugging a 5-inch glass pipette with a small plug of glass wool. Use a long pipette to tap the glass wool down into the neck of the pipette. Add 2 cm of anhydrous sodium sulfate and 2 cm of alumina.
- ❑ Pass the organic layer remaining in the reaction vessel through the column of alumina and sodium sulfate into a tared 25-mL round-bottom flask.
- ❑ Rinse the column with diethyl ether (2 mL) into the round-bottom flask.
- ❑ Remove the diethyl ether from your organic solution (either under reduced pressure or using a stream of nitrogen gas) until a constant weight is observed.
- ❑ Re-weigh the flask containing the product and calculate the yield.

## Characterize the Product

- ❑ To determine the purity of the product, run a TLC of a sample using 20% ethyl acetate in hexane as eluent.
- ❑ Obtain an IR spectrum of the product and compare it with that of the starting material.
- ❑ Obtain $^{1}H$-NMR and $^{13}C$-NMR spectra in $CDCl_3$ if instructed to do so.

## PROCEDURE FOR USE IN A MULTIMODE MICROWAVE UNIT

$Na_2WO_4$, $H_2O_2$

$(C_9H_1)_3NHSO_4$

Your laboratory instructor will assign you an alcohol to use in this reaction. Prepare your notebook by writing your pre-lab assignment incorporating your assigned alcohol.

**Caution:** Sodium tungstate, methyltrioctylammonium hydrogen sulfate, and 1-phenyl-1-propanol are irritants so avoid skin contact. Aqueous 30% hydrogen peroxide is very corrosive and a strong oxidant. Avoid contact of the reagent with the skin and any metal objects such as spatulas. Transfer the hydrogen peroxide solution to a glass graduated cylinder and pipette the required amount directly into the reaction tube. Flush any extra reagent down the sink with large amounts of water. If any peroxide is accidentally spilled on the skin, immediately wash the area with large amounts of water to prevent a burn. The use of goggles with side-shields, lab coats, and gloves is considered minimum and nondiscretionary safety practice in the laboratory.

### Table of Reagents and Physical Constants

| Reagent | Equiv. | FW | mmol | Mass (mg) | Density (g/mL) | Vol. (mL) | mp/bp°C |
|---|---|---|---|---|---|---|---|
| Alcohol ROH | 1 | — | 4.2 | — | — | — | — |
| Sodium tungstate dihydrate $Na_2WO_4{\cdot}H_2O$ | 0.01 | 330 | 0.042 | 14 | — | — | 698 |
| Methyltrioctyl ammonium hydrogen sulfate $C_{67}H_{76}NSO_4$ | 0.01 | 466 | 0.042 | 20 | — | — | 61–65 |
| Hydrogen peroxide (30% aqueous) $H_2O_2$ | — | 34.0 | — | — | — | 4.0 | — |
| ***Alcohols*** | | | | | | | |
| 1-Phenyl-1-propanol $C_9H_{12}O$ | 1 | 136 | 4.2 | 572 | 0.994 | 0.58 | 219 |
| 1-Phenyl-1-ethanol $C_8H_{10}O$ | 1 | 122 | 4.2 | 513 | 1.012 | 0.51 | 19–20 |
| 4-*tert*-Butyl cyclohexanol $C_{10}H_{20}O$ | 1 | 156 | 4.2 | 660 | — | — | 62–70 |

## Perform the Reaction

- ❑ In a 25-mL glass microwave reaction vessel place a magnetic stir bar, the alcohol, sodium tungstate dihydrate, and methyltrioctylammonium hydrogen sulfate.
- ❑ Using a glass graduated cylinder add the 30% hydrogen peroxide to the reaction vessel.
- ❑ Seal the reaction vessel with a cap according to the microwave manufacturer's recommendations.
- ❑ Carefully stir the contents on a magnetic stir plate to dissolve the solid.
- ❑ Place the sealed reaction vessel in the carousel, noting the vessel's position number and ensuring that vessels are evenly spaced around the carousel.
- ❑ When all the group's reaction vessels are in place, load the carousel into the microwave cavity.
- ❑ If provided by the manufacturer, connect the temperature probe to the control vessel.
- ❑ Program the microwave unit to heat the contents of the reaction vessels to 150°C over a 2-minute ramp period and then hold at this temperature for 4 minutes.
- ❑ After the heating step is complete, allow the contents of the reaction vessel to cool to 50°C or below before removing it from the microwave cavity.

## Isolate the Product

- ❑ Carefully open the reaction vessel.
- ❑ Add water (4 mL) to the contents of the reaction vessel, followed by diethyl ether (8 mL).
- ❑ Agitate the vessel contents by placing the tube on a magnetic stir plate.
- ❑ Clamp the vessel and allow the contents to separate into two layers.
- ❑ Remove the bottom aqueous layer with a pipette and place the aqueous layer in a 25-mL Erlenmeyer flask.
- ❑ Add a saturated solution of sodium sulfite (8 mL) to the organic layer in the reaction vessel.
- ❑ Agitate the contents of the vessel again.
- ❑ Allow the layers to separate and remove the bottom aqueous layer as before.
- ❑ With a magnetic retrieving wand remove the stir bar.
- ❑ Prepare a "drying pipette" by plugging a 5-inch glass pipette with a small plug of glass wool. Use a long pipette to tap the glass wool down into the neck of the pipette. Add 2 cm of anhydrous sodium sulfate and 2 cm of alumina.
- ❑ Pass the organic layer remaining in the reaction vessel through the column of alumina and sodium sulfate into a tared 25-mL round-bottom flask.
- ❑ Rinse the column with diethyl ether (4 mL) into the round-bottom flask.
- ❑ Remove the diethyl ether from the organic solution (either under reduced pressure or using a stream of nitrogen gas) until a constant weight is observed.
- ❑ Re-weigh the flask containing the product and calculate the yield.

## Characterize the Product

- ❑ To determine the purity of the product, run a TLC of a sample using 20% ethyl acetate in hexane as eluent.
- ❑ Obtain an IR spectrum of the product and compare it with that of the starting material.
- ❑ Obtain $^{1}$H-NMR and $^{13}$C-NMR spectra in $CDCl_3$ if instructed to do so.

## QUESTIONS

1. Are the following transformations oxidation reactions according to the definition that oxidation is the "addition of more oxygen bonds to carbon; the addition of more electronegative atoms bonded to carbon; the loss of hydrogen atoms bonded to carbon"?

(a)

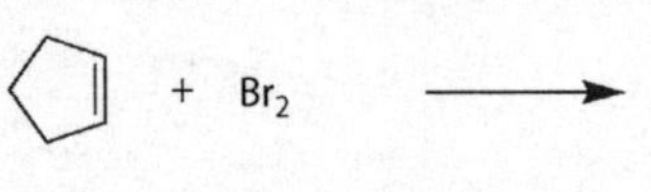


(b)

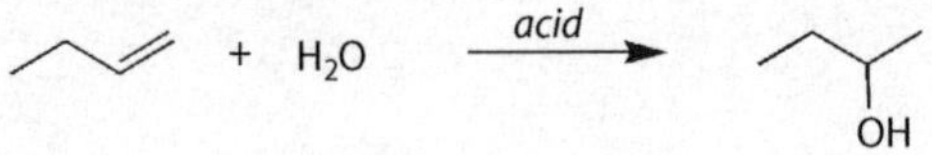


(c)

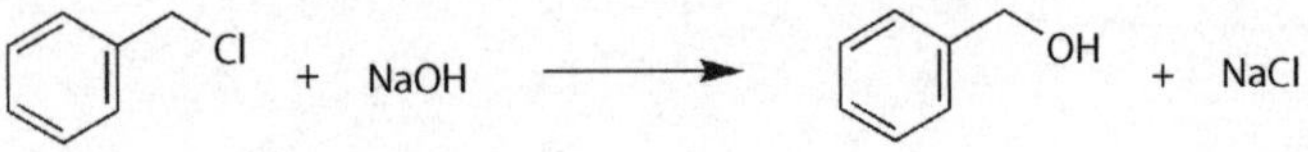


(d)

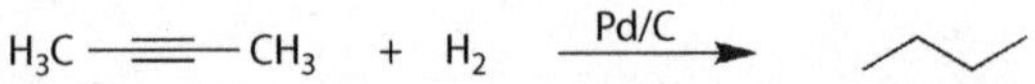


(e)

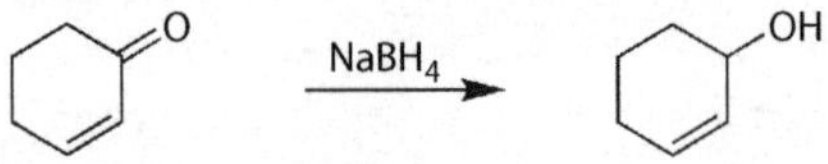


2. What is the role of a phase-transfer catalyst in a reaction?
3. Which compound would you predict to have the higher $R_f$ on the TLC plate, the starting alcohol or the product ketone?
4. What major absorptions in the infrared spectrum do you observe that would lead you to believe an oxidation of an alcohol to a ketone had indeed taken place?

# 12 Suzuki Coupling Reaction
## *Preparation of a Biaryl*

**LEARNING GOALS**

- To perform a metal-catalyzed reaction
- To use water as a solvent for organic synthesis

## INTRODUCTION

Metal-catalyzed reactions are used widely for making important organic molecules. This was reflected in the 2010 Nobel Prize in Chemistry which was awarded to three chemists for their pioneering work in this field, specifically the development of palladium-catalyzed cross-coupling reactions. An example of this is the Suzuki coupling reaction of an aryl halide with a boronic acid using a palladium catalyst to give a biaryl, the reaction being named after its pioneer.

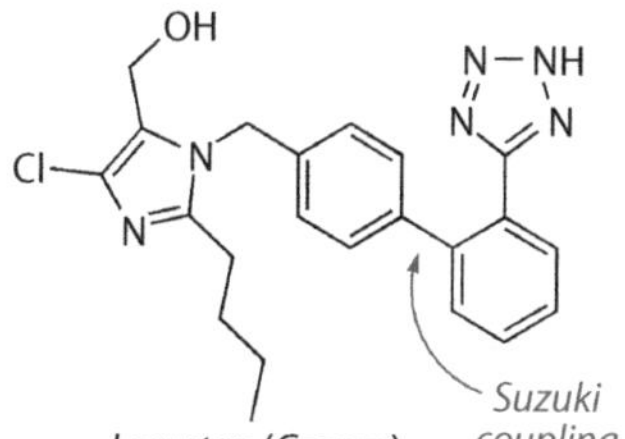

*Losartan* (*Cozaar*) — *Suzuki coupling*

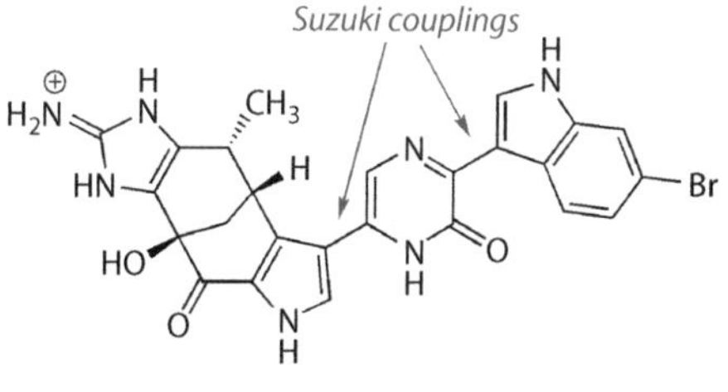

(+)-*Dragmacidin F* — *Suzuki couplings*

The Suzuki reaction is used industrially for the manufacture of pharmaceuticals and natural products. For example, the drug Losartan (used to treat high blood pressure) is prepared using a Suzuki coupling as one of the key steps. The same is true for dragmacidin F, a bromoindole alkaloid with antiviral properties, that was isolated from a marine sponge discovered off the coast of Italy. In this case, each sponge contains only a small amount of the active compound and it is both unrealistic and bad practice to harvest lots of sponges to extract enough of it to be tested in the clinic. However, by using the Suzuki coupling reaction, it is possible to make the compound in the lab and its properties can now be more thoroughly tested.

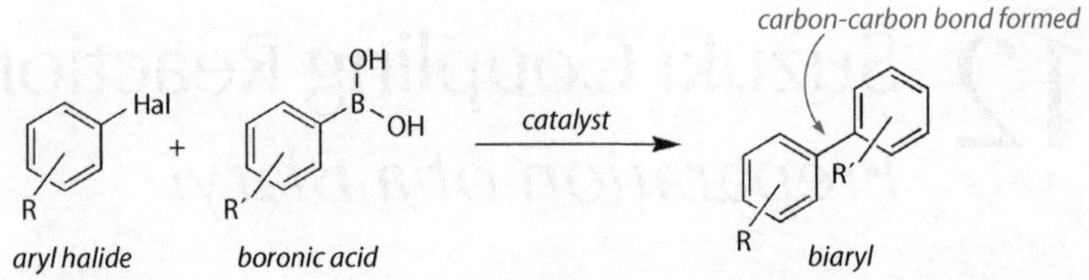

The palladium catalysts used for the reactions generally have groups around them called ligands (L) and the general formula of the catalyst is often written as $PdL_n$. The reaction occurs through a catalytic cycle in which first the palladium inserts into the carbon–halogen bond of the aryl halide. Then, after reaction with the base, the ring from the boronic acid is also attached to the palladium atom. In the final step of the cycle, the palladium catalyst is ejected as the biaryl is formed. The palladium catalyst can then start the cycle again.

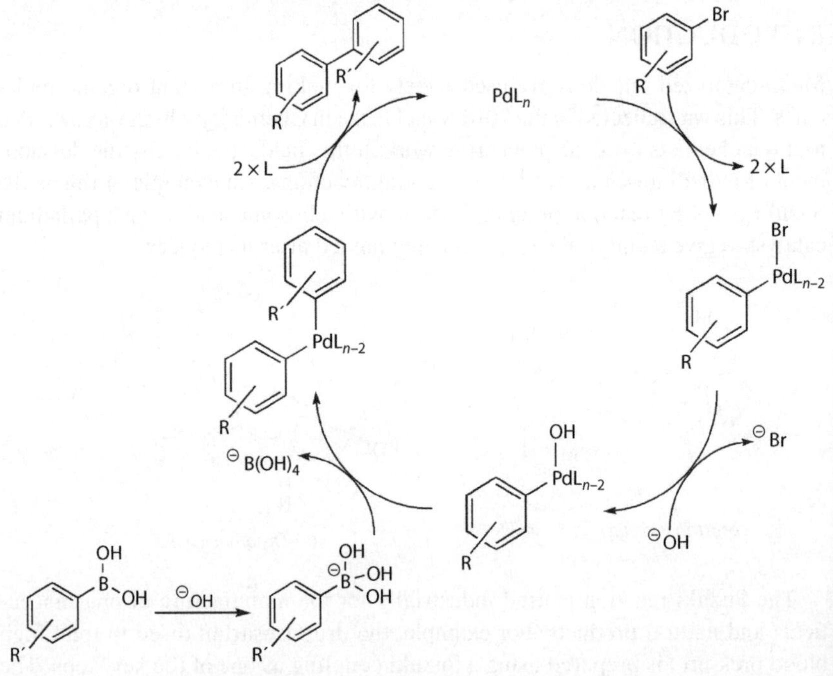

A wide range of catalysts has been developed for the reaction, many of them being quite costly and hard to remove from the product. In this experiment, cheap, readily available palladium chloride, $PdCl_2$, is used as the catalyst and the reaction is run in water as a solvent. Although most organic compounds are not soluble in water at room temperature, they can be soluble, or at least partly so, in hot water. A co-solvent such as ethanol or a phase-transfer catalyst such as tetrabutylammonium bromide ($Bu_4NBr$) can also be used to facilitate the dissolution of organic substrates in water.

# PROCEDURE FOR USE IN A MONOMODE MICROWAVE UNIT

## PREPARATION OF 4-ACETYLBIPHENYL

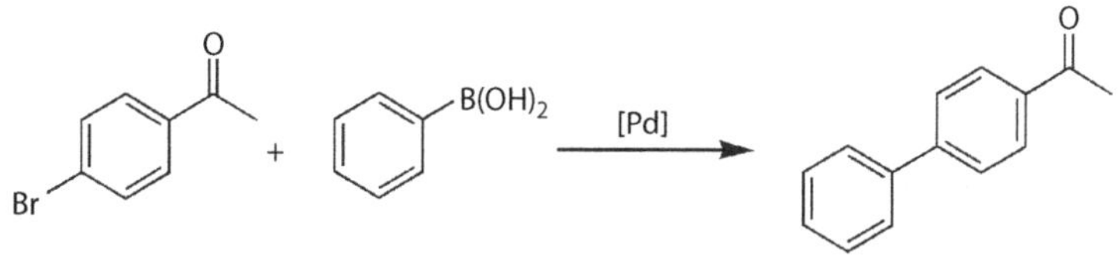

**Caution:** Sodium carbonate, phenylboronic acid, and 4′-bromoacetophenone are irritants. The palladium stock solution is toxic. The use of goggles with side-shields, lab coats, and gloves is considered minimum and nondiscretionary safety practice in the laboratory.

### Table of Reagents and Physical Constants

| Reagent | Equiv. | FW | mmol | Mass (mg) | Vol. (mL) | mp°C |
|---|---|---|---|---|---|---|
| 4-Bromoacetophenone $C_8H_7BrO$ | 1.0 | 199 | 1.0 | 199 | — | 49–51 |
| Phenylboronic acid $C_6H_7BO_2$ | 1.2 | 122 | 1.2 | 146 | — | 217–222 |
| Sodium carbonate $Na_2CO_3$ | 3.0 | 106 | 3.0 | 318 | — | 851 |
| Palladium chloride (1000 ppm stock solution) $PdCl_2$ | — | — | — | — | 0.4 | — |
| Ethanol $C_2H_6O$ | — | — | — | — | 1.0 | 78 |
| Water $H_2O$ | — | — | — | — | 1.0 | 100 |

## PERFORM THE REACTION

- ❑ In a 10-mL glass microwave reaction vessel, place a magnetic stir bar, the 4-bromoacetophenone, phenylboronic acid, and sodium carbonate.
- ❑ Using a graduated cylinder, add the water and ethanol to the reaction vessel.
- ❑ Using an automatic delivery pipette, add the palladium stock solution to the reaction vessel.
- ❑ Seal the reaction vessel with a cap according to the microwave manufacturer's recommendations.
- ❑ Place the sealed reaction vessel in the microwave cavity.
- ❑ Program the microwave unit to heat the vessel contents to 150°C over a 3-minute ramp period and then hold at this temperature for 5 minutes.

- ❑ After the heating step is completed, allow the contents of the reaction vessel to cool to 50°C or below before removing it from the microwave cavity.

## Isolate the Product

- ❑ While the reaction is cooling, obtain two 100-mL Erlenmeyer flasks and label them “aqueous phase” and “organic phase”.
- ❑ Carefully open the cooled reaction vessel.
- ❑ Transfer the contents of the reaction vessel to a 125-mL separatory funnel.
- ❑ Rinse the reaction vessel with diethyl ether (3 mL).
- ❑ Transfer the diethyl ether washings to the separatory funnel.
- ❑ Add a further 7 mL of diethyl ether and 10 mL of water to the separatory funnel.
- ❑ Carefully stopper the separatory funnel and invert the funnel.
- ❑ Immediately vent the funnel by opening the stopcock to release pressure that may have developed.
- ❑ Close the stopcock and mix the two layers several times by inverting the funnel repeatedly.
- ❑ Vent the funnel.
- ❑ Close the stopcock and re-clamp the funnel to a ring stand and remove the stopper.
- ❑ Allow the layers to separate.
- ❑ Move the bottom aqueous layer through the stopcock into the 100-mL Erlenmeyer flask labeled “aqueous phase”.
- ❑ Pour the ether layer through the top of the separatory funnel into the 100-mL Erlenmeyer flask labeled “organic phase”.
- ❑ Return the aqueous layer to the separatory funnel.
- ❑ Extract the aqueous layer by adding diethyl ether (10 mL).
- ❑ Carefully stopper the separatory funnel and invert the funnel.
- ❑ Immediately vent the funnel by opening the stopcock to reduce pressure that may have developed in the funnel.
- ❑ Close the stopcock and mix the two layers several times by inverting the funnel repeatedly.
- ❑ Vent the funnel.
- ❑ Close the stopcock and re-clamp the funnel to a ring stand and remove the stopper.
- ❑ Allow the layers to separate.
- ❑ Move the bottom aqueous layer through the stopcock into the 100-mL Erlenmeyer flask labeled “aqueous phase”.
- ❑ Pour the ether layer through the top of the separatory funnel into the 100-mL Erlenmeyer flask labeled “organic phase” that already contains the previous ether layer.
- ❑ Return the aqueous layer to the separatory funnel.
- ❑ Repeat the extraction of the aqueous layer with 10-mL diethyl ether collecting the ether layer in the Erlenmeyer flask labeled “organic phase”.

- ❏ Rinse the separatory funnel with water and acetone discarding the rinses in appropriate waste containers.
- ❏ Clamp the separatory funnel to the ring stand and close the stopcock.
- ❏ Return to the separatory funnel the combined ether layers from the Erlenmeyer flask labeled "organic phase" (approximately 30 mL).
- ❏ Add 5% sodium bicarbonate solution to the separatory funnel.
- ❏ Carefully stopper the separatory funnel and invert the funnel.
- ❏ Immediately vent the funnel by opening the stopcock to reduce pressure that may have developed in the funnel.
- ❏ Close the stopcock and mix the two layers several times by inverting the funnel repeatedly.
- ❏ Vent the funnel.
- ❏ Close the stopcock and re-clamp the funnel to a ring stand and remove the stopper.
- ❏ Allow the layers to separate.
- ❏ Move the bottom aqueous layer through the stopcock into the 100-mL Erlenmeyer flask labeled "aqueous phase".
- ❏ Add saturated sodium chloride (30 mL) to the separatory funnel.
- ❏ Carefully stopper the separatory funnel and invert the funnel.
- ❏ Immediately vent the funnel by opening the stopcock to reduce pressure that may have developed in the funnel.
- ❏ Close the stopcock and mix the two layers several times by inverting the funnel repeatedly.
- ❏ Vent the funnel.
- ❏ Close the stopcock and re-clamp the funnel to a ring stand and remove the stopper.
- ❏ Allow the layers to separate.
- ❏ Move the bottom aqueous layer through the stopcock into the 100-mL Erlenmeyer flask labeled "aqueous phase".
- ❏ Pour the organic phase into the clean 50-mL Erlenmeyer flask.
- ❏ Dry this organic layer by adding to the flask containing the ether layer a drying reagent, either anhydrous magnesium sulfate ($MgSO_4$) or anhydrous sodium sulfate ($Na_2SO_4$).
- ❏ Decant the diethyl ether solution into a 100-mL round-bottom flask leaving the drying agent behind.
- ❏ Rinse the Erlenmeyer flask and drying reagent by adding to the flask diethyl ether (3 mL).
- ❏ Swirl the flask.
- ❏ Decant the ether rinse into the round-bottom flask.
- ❏ Remove the ether solvent under reduced pressure until a constant weight is reached.
- ❏ Scrape out the white solid, placing it into a tared vial.
- ❏ Re-weigh the vial plus contents.
- ❏ Calculate the yield and percent yield.

## Characterize the Product

- ❑ To determine the purity of the product, obtain a melting point.
- ❑ Obtain an IR spectrum of the product and compare it with that of the starting material.
- ❑ Obtain $^{1}$H-NMR and $^{13}$C-NMR spectra in $CDCl_3$ if instructed to do so.

# PROCEDURE FOR USE IN A MULTIMODE MICROWAVE UNIT

## PREPARATION OF 4-ACETYLBIPHENYL

Br–C₆H₄–C(O)CH₃ + C₆H₅B(OH)₂ —[Pd]→ C₆H₅–C₆H₄–C(O)CH₃

**Caution:** Sodium carbonate, phenylboronic acid, and 4′-bromoacetophenone are irritants. The palladium stock solution is toxic. The use of goggles with side-shields, lab coats, and gloves is considered minimum and nondiscretionary safety practice in the laboratory.

### Table of Reagents and Physical Constants

| Reagent | Equiv. | FW | mmol | Mass (mg) | Vol. (mL) | mp°C |
|---|---|---|---|---|---|---|
| 4-Bromoacetophenone $C_8H_7BrO$ | 1.0 | 199 | 2.0 | 398 | — | 49–51 |
| Phenylboronic acid $C_6H_7BO_2$ | 1.2 | 122 | 2.4 | 292 | — | 217–222 |
| Sodium carbonate $Na_2CO_3$ | 3.0 | 106 | 6.0 | 636 | — | 851 |
| Palladium chloride (1000 ppm stock solution) $PdCl_2$ | — | — | — | — | 0.8 | — |
| Ethanol $C_2H_6O$ | — | — | — | — | 3.0 | 78 |
| Water $H_2O$ | — | — | — | — | 3.0 | 100 |

## PERFORM THE REACTION

- ❑ In a 25-mL glass microwave reaction vessel, place a magnetic stir bar, the 4′-bromoacetophenone, phenylboronic acid, and sodium carbonate.
- ❑ Using a graduated cylinder, add the water and ethanol to the reaction vessel.
- ❑ Using an automatic delivery pipette, add the palladium stock solution to the reaction vessel.
- ❑ Seal the reaction vessel with a cap according to the microwave manufacturer's recommendations.
- ❑ Place the sealed reaction vessel in the carousel, noting the vessel's position number and ensuring that vessels are evenly spaced around the carousel.
- ❑ When all the group's reaction vessels are in place, load the carousel into the microwave cavity.

- ❑ If provided by the manufacturer, connect the temperature probe to the control vessel.
- ❑ Program the microwave unit to heat the vessel contents to 150°C over a 10-minute ramp period and then hold at this temperature for 5 minutes.
- ❑ After the heating step is completed, allow the contents of the reaction vessel to cool to 50°C or below before removing it from the microwave cavity.

## Isolate the Product

- ❑ While the reaction is cooling, obtain two 100-mL Erlenmeyer flasks and label them "aqueous phase" and "organic phase".
- ❑ Carefully open the cooled reaction vessel.
- ❑ Transfer the contents of the reaction vessel to a 125-mL separatory funnel.
- ❑ Rinse the reaction vessel with diethyl ether (3 mL).
- ❑ Transfer the diethyl ether washings to the separatory funnel.
- ❑ Add a further 7 mL of diethyl ether and 10 mL of water to the separatory funnel.
- ❑ Carefully stopper the separatory funnel and invert the funnel.
- ❑ Immediately vent the funnel by opening the stopcock to release pressure that may have developed.
- ❑ Close the stopcock and mix the two layers several times by inverting the funnel repeatedly.
- ❑ Vent the funnel.
- ❑ Close the stopcock and re-clamp the funnel to a ring stand and remove the stopper.
- ❑ Allow the layers to separate.
- ❑ Move the bottom aqueous layer through the stopcock into the 100-mL Erlenmeyer flask labeled "aqueous phase".
- ❑ Pour the ether layer through the top of the separatory funnel into the 100-mL Erlenmeyer flask labeled "organic phase".
- ❑ Return the aqueous layer to the separatory funnel.
- ❑ Extract the aqueous layer by adding diethyl ether (10 mL).
- ❑ Carefully stopper the separatory funnel and invert the funnel.
- ❑ Immediately vent the funnel by opening the stopcock to reduce pressure that may have developed in the funnel.
- ❑ Close the stopcock and mix the two layers several times by inverting the funnel repeatedly.
- ❑ Vent the funnel.
- ❑ Close the stopcock and re-clamp the funnel to a ring stand and remove the stopper.
- ❑ Allow the layers to separate.
- ❑ Move the bottom aqueous layer through the stopcock into the 100-mL Erlenmeyer flask labeled "aqueous phase".

- ❑ Pour the ether layer through the top of the separatory funnel into the 100-mL Erlenmeyer flask labeled "organic phase" that already contains the previous ether layer.
- ❑ Return the aqueous layer to the separatory funnel.
- ❑ Repeat the extraction of the aqueous layer with 10 mL diethyl ether collecting the ether layer in the Erlenmeyer flask labeled "organic phase".
- ❑ Rinse the separatory funnel with water and acetone, discarding the rinses in appropriate waste containers.
- ❑ Clamp the separatory funnel to the ring stand and close the stopcock.
- ❑ Return to the separatory funnel the combined ether layers from the Erlenmeyer flask labeled "organic phase" (approximately 30 mL).
- ❑ Add 5% sodium bicarbonate solution to the separatory funnel.
- ❑ Carefully stopper the separatory funnel and invert the funnel.
- ❑ Immediately vent the funnel by opening the stopcock to reduce pressure that may have developed in the funnel.
- ❑ Close the stopcock and mix the two layers several times by inverting the funnel repeatedly.
- ❑ Vent the funnel.
- ❑ Close the stopcock and re-clamp the funnel to a ring stand and remove the stopper.
- ❑ Allow the layers to separate.
- ❑ Remove the bottom aqueous layer through the stopcock into the 100-mL Erlenmeyer flask labeled "aqueous phase".
- ❑ Add saturated sodium chloride (30 mL) to the separatory funnel.
- ❑ Carefully stopper the separatory funnel and invert the funnel.
- ❑ Immediately vent the funnel by opening the stopcock to reduce pressure that may have developed in the funnel.
- ❑ Close the stopcock and mix the two layers several times by inverting the funnel repeatedly.
- ❑ Vent the funnel.
- ❑ Close the stopcock and re-clamp the funnel to a ring stand and remove the stopper.
- ❑ Allow the layers to separate.
- ❑ Move the bottom aqueous layer through the stopcock into the 100-mL Erlenmeyer flask labeled "aqueous phase".
- ❑ Pour the organic phase into the clean 50-mL Erlenmeyer flask.
- ❑ Dry this organic layer by adding to the flask containing the ether layer a drying reagent, either anhydrous magnesium sulfate ($MgSO_4$) or anhydrous sodium sulfate ($Na_2SO_4$).
- ❑ Decant the diethyl ether solution into a 100-mL round-bottom flask leaving the drying agent behind.
- ❑ Rinse the Erlenmeyer flask and drying reagent by adding to the flask diethyl ether (3 mL).
- ❑ Swirl the flask.
- ❑ Decant the ether rinse into the round-bottom flask.

- ❑ Remove the ether solvent under reduced pressure until a constant weight is reached.
- ❑ Scrape out the white solid, placing it into a tared vial.
- ❑ Re-weigh the vial plus contents.
- ❑ Calculate the yield and percent yield.

## Characterize the Product

- ❑ To determine the purity of the product, obtain a melting point.
- ❑ Obtain an IR spectrum of the product and compare it with that of the starting material.
- ❑ Obtain $^{1}H$-NMR and $^{13}C$-NMR spectra in $CDCl_3$ if instructed to do so.

## QUESTIONS

1. What is the palladium catalyst loading in your reaction (in mol%)?
2. What two purposes does the sodium carbonate serve during the reaction?
3. What is the purpose of washing with sodium bicarbonate? What does it remove?
4. What other palladium-catalyzed cross-coupling reactions were recognized in the 2010 Nobel Prize for Chemistry?

# 13 Heck Reaction
## *Preparation of Substituted Cinnamic Acids*[*]

**LEARNING GOALS**

- To perform a metal-catalyzed coupling reaction
- To separate a solid product from a solution by vacuum filtration

## INTRODUCTION

Palladium-catalyzed couplings are some of the most widely used carbon–carbon bond-forming reactions in organic synthesis. Two examples that are covered in most organic textbooks are the Suzuki reaction and the Heck reaction. These involve palladium(0) complexes, which are often generated in situ by reduction of palladium(II) precursors such as palladium acetate. In the Suzuki reaction, usually an aryl or alkyl halide is coupled with an aryl or alkyl boronic acid. In the Heck reaction, an aryl or alkyl halide is coupled with an alkene.

*Suzuki Reaction* $R{-}X + R'{-}B(OH)_2 \xrightarrow{Pd(0)} R{-}R'$

*Heck Reaction* $R{-}X + R'{-}CH{=}CH_2 \xrightarrow{Pd(0)} R'{-}CH{=}CH{-}R$

*R and R′ = alkyl or aryl;   X = halogen*

The first step in the catalytic cycle for the Heck reaction involves insertion of palladium(0) into the carbon–halogen bond of the halogenated reagent in an oxidative addition.

$L_2Pd(0) + R{-}X \longrightarrow L_2Pd(II)(R)(X)$

[*] Experimental procedure developed by Michael Luciano, Merrimack College.

This complex then undergoes a *syn* addition with the alkene component to form a carbon–carbon single bond between the reagents. This then rotates internally to relieve steric strain.

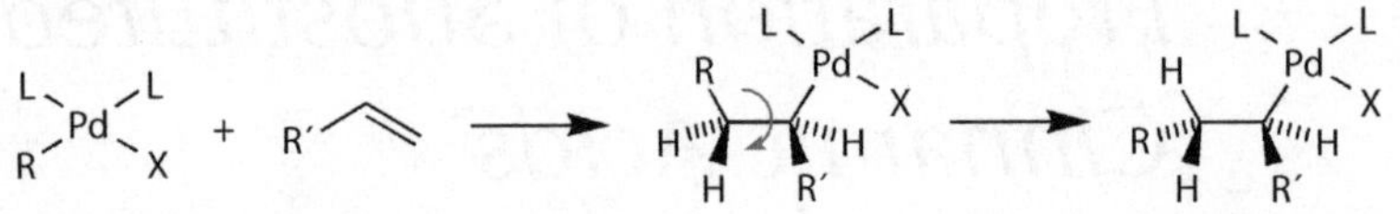

The carbon–carbon double bond is then formed through a *syn*-β-elimination in the more stable Z-conformation.

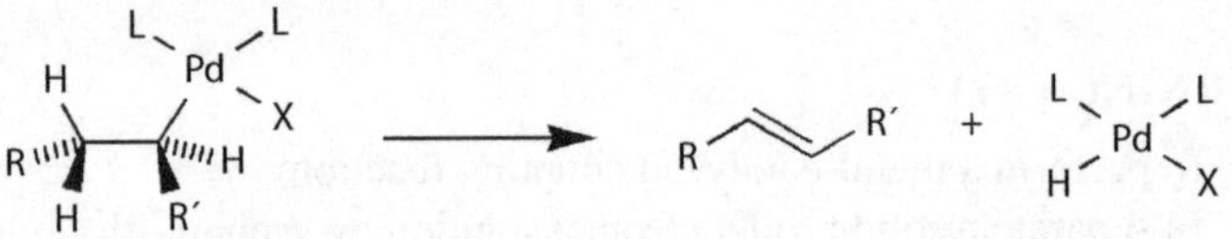

The palladium complex then undergoes a reductive elimination of HX to regenerate the palladium(0) species.

Methyl acrylate or acrylic acid can couple with an aryl halide to form a β-substituted cinnamic acid. The intermediate product formed by the Heck reaction when using the methyl acrylate is the methyl ester of cinnamic acid. The ester is subsequently hydrolyzed under the basic aqueous conditions used in the reaction, generating the corresponding carboxylic acid.

## PROCEDURE FOR USE IN A MONOMODE MICROWAVE UNIT

### PREPARATION OF 4-METHOXYCINNAMIC ACID (2*E*-3-(4-METHOXYPHENYL)-2-PROPENOIC ACID)

$$\text{4-}H_3CO\text{-}C_6H_4\text{-Br} + CH_2{=}CH\text{-}C(O)OCH_3 \xrightarrow[K_2CO_3,\ H_2O]{[Pd]} \text{4-}H_3CO\text{-}C_6H_4\text{-}CH{=}CH\text{-}C(O)OH$$

**Caution:** 4-Bromoanisole and methyl acrylate are lachrymatory and irritants. They should be measured out in a fume cupboard or well-ventilated area. The palladium stock solution is toxic. The use of goggles with side-shields, lab coats, and gloves is considered minimum and nondiscretionary safety practice in the laboratory.

### Table of Reagents and Physical Constants

| Reagent | Equiv. | FW | mmol | Mass (mg) | Density (g/mL) | Vol. (mL) | mp/bp°C |
|---|---|---|---|---|---|---|---|
| 4-Bromoanisole $C_7H_7BrO$ | 1 | 187 | 1.0 | 187 | 1.49 | 0.126 | 223 |
| Methyl acrylate $C_4H_6O_2$ | 2 | 86.0 | 2.0 | 172 | 0.95 | 0.181 | 80 |
| Palladium chloride (1,000 ppm stock solution) $PdCl_2$ | — | — | — | — | — | 0.4 | — |
| Tetrabutylammonium bromide $C_{16}H_{36}BrN$ | 1 | 322 | 1.0 | 322 | — | — | 103 |
| Potassium carbonate $K_2CO_3$ | 3.7 | 138 | 3.7 | 511 | — | — | — |
| Water $H_2O$ | — | — | — | — | — | 1.6 | 100 |

### PERFORM THE REACTION

- ❑ In a 10-mL glass microwave reaction vessel place a magnetic stir bar, the tetrabutylammonium bromide, and the potassium carbonate.
- ❑ Using a graduated cylinder, add the water to the reaction vessel.
- ❑ Using an automatic delivery pipette, add the 4-bromoanisole, methyl acrylate, and palladium stock solution to the reaction vessel.
- ❑ Seal the reaction vessel with a cap according to the microwave manufacturer's recommendations.

- ❑ Program the microwave unit to heat the vessel contents to 175°C using an initial microwave power of 250 W and hold at this temperature for 20 minutes.
- ❑ After the heating step is completed, allow the contents of the reaction vessel to cool to 50°C or below before removing it from the microwave cavity.

## Isolate the Product

- ❑ Add water (4 mL) to a clean 25-mL Erlenmeyer flask.
- ❑ Carefully open the reaction vessel and pipette the contents of the reaction vessel into the Erlenmeyer flask containing the water.
- ❑ Add 6 M HCl dropwise to the Erlenmeyer flask until the pH reaches 6. The product will precipitate out of the solution.
- ❑ Cool the solution in the ice bath for 10 minutes after the addition is complete.
- ❑ While the solution is cooling, set up a vacuum-filtration apparatus with a Hirsch funnel, side-arm flask, rubber collar, and a length of rubber vacuum tubing.
- ❑ Cool some distilled water (10 mL) in a 25-mL Erlenmeyer flask by placing it in an ice bath for washing the product after filtration.
- ❑ Connect the filtration system to a vacuum and place the correct size filter paper in the funnel.
- ❑ Wet the filter paper with a few drops of the cold distilled water and start the vacuum to seal the filter paper in place.
- ❑ Filter the reaction mixture by pouring the contents of the 25-mL Erlenmeyer flask into the funnel; transfer as much solid as possible.
- ❑ Rinse the flask with cold distilled water (3 mL) and add it to the filter funnel.
- ❑ Rinse the precipitate with additional cold distilled water (3 mL).
- ❑ Allow the precipitate to dry on the funnel for 10 minutes.
- ❑ Transfer the solid to a large piece of filter paper to dry completely.
- ❑ Weigh the dry product and calculate the yield and percent yield.

## Characterize the Product

- ❑ Determine the melting point of the white product.
- ❑ If the melting point indicates the product is not pure, re-crystallize the crude product from ethanol.
- ❑ Obtain an IR spectrum of the product and compare it with that of the starting material.
- ❑ Obtain $^1$H-NMR and $^{13}$C-NMR spectra in $CDCl_3$ if instructed to do so.

# PROCEDURE FOR USE IN A MULTIMODE MICROWAVE UNIT

## Preparation of 4-Methoxycinnamic Acid (2*E*-3-(4-Methoxyphenyl)-2-Propenoic Acid)

[4-bromoanisole ($H_3CO$, Br) + methyl acrylate ($OCH_3$) → [Pd], $K_2CO_3$, $H_2O$ → 4-methoxycinnamic acid ($H_3CO$, OH)]

**Caution:** 4-Bromoanisole and methyl acrylate are lachrymatory and irritants. They should be measured out in a fume cupboard or well-ventilated area. The palladium stock solution is toxic. The use of goggles with side-shields, lab coats, and gloves is considered minimum and nondiscretionary safety practice in the laboratory.

### Table of Reagents and Physical Constants

| Reagent | Equiv. | FW | mmol | Mass (mg) | Density (g/mL) | Vol. (mL) | mp/bp°C |
|---|---|---|---|---|---|---|---|
| 4-Bromoanisole $C_7H_7BrO$ | 1 | 187 | 1.0 | 187 | 1.49 | 0.126 | 223 |
| Methyl acrylate $C_4H_6O_2$ | 2 | 86.0 | 2.0 | 172 | 0.95 | 0.187 | 80 |
| Palladium chloride (1,000 ppm stock solution) $PdCl_2$ | — | — | — | — | — | 0.4 | — |
| Tetrabutylammonium bromide $C_{16}H_{36}BrN$ | 1 | 322 | 1.0 | 322 | — | — | 103 |
| Potassium carbonate $K_2CO_3$ | 3.7 | 138 | 3.7 | 511 | — | — | — |
| Water $H_2O$ | — | — | — | — | — | 4.0 | 100 |

### Perform the Reaction

- ❑ In a 25-mL glass microwave reaction vessel place a magnetic stir bar, the tetrabutylammonium bromide, and the potassium carbonate.
- ❑ Using a graduated cylinder add the water to the reaction vessel.
- ❑ Using an automatic delivery pipette, add the 4-bromoanisole, methyl acrylate, and palladium stock solution to the reaction vessel.
- ❑ Seal the reaction vessel with a cap according to the microwave manufacturer's recommendations.
- ❑ Place the sealed reaction vessel in the carousel, noting the vessel's position number and ensuring that vessels are evenly spaced around the carousel.

- ❑ When all the group's reaction vessels are in place, load the carousel into the microwave cavity.
- ❑ If provided by the manufacturer, connect the temperature probe to the control vessel.
- ❑ Program the microwave unit to heat the vessel contents to 175°C using an initial microwave power of 600 W and hold at this temperature for 20 minutes.
- ❑ After the heating step is completed, allow the contents of the reaction vessel to cool to 50°C or below before removing it from the microwave cavity.

## Isolate the Product

- ❑ Add water (4 mL) to a clean 25-mL Erlenmeyer flask.
- ❑ Carefully open the reaction vessel and pipette the contents of the reaction vessel into the Erlenmeyer flask containing the water.
- ❑ Add 6 M HCl dropwise to the Erlenmeyer flask until the pH reaches 6. The product will precipitate out of the solution.
- ❑ Cool the solution in the ice bath for 10 minutes after the addition is complete.
- ❑ While the solution is cooling, set up a vacuum-filtration apparatus with a Hirsch funnel, side-arm flask, rubber collar, and a length of rubber vacuum tubing.
- ❑ Cool some distilled water (10 mL) in a 25-mL Erlenmeyer flask by placing it in an ice bath for washing the product after filtration.
- ❑ Connect the filtration system to a vacuum and place the correct size filter paper in the funnel.
- ❑ Wet the filter paper with a few drops of the cold distilled water and start the vacuum to seal the filter paper in place.
- ❑ Filter the reaction mixture by pouring the contents of the 25-mL Erlenmeyer flask into the funnel; transfer as much solid as possible.
- ❑ Rinse the flask with cold distilled water (3 mL) and add it to the filter funnel.
- ❑ Rinse the precipitate with additional cold distilled water (3 mL).
- ❑ Allow the precipitate to dry on the funnel for 10 minutes.
- ❑ Transfer the solid to a large piece of filter paper to dry completely.
- ❑ Weigh the dry product and calculate the yield and percent yield.

## Characterize the Product

- ❑ Determine the melting point of the white product.
- ❑ If the melting point indicates the product is not pure, re-crystallize the crude product from ethanol.
- ❑ Obtain an IR spectrum of the product and compare it with that of the starting material.
- ❑ Obtain $^{1}$H-NMR and $^{13}$C-NMR spectra in $CDCl_3$ if instructed to do so.

## PROCEDURE FOR USE IN A MONOMODE MICROWAVE UNIT

### SYNTHESIS OF 4-ACETYLCINNAMIC ACID (2*E*-3-(4-ACETYLPHENYL)-2-PROPENOIC ACID)

[Pd]

$K_2CO_3$, $H_2O$

**Caution:** Acrylic acid is lachrymatory and an irritant. It should be measured out in a fume hood or well-ventilated area. 4-Iodoacetophenone is an irritant. The palladium stock solution is toxic. The use of goggles with side-shields, lab coats, and gloves is considered minimum and nondiscretionary safety practice in the laboratory.

**Table of Reagents and Physical Constants**

| Reagent | Equiv. | FW | mmol | Mass (mg) | Density (g/mL) | Vol. (mL) | mp/bp°C |
|---|---|---|---|---|---|---|---|
| 4-Iodoacetophenone $C_8H_7IO$ | 1 | 246 | 1.0 | 246 | — | — | 82–84 |
| Acrylic acid $C_3H_4O_2$ | 2 | 72.0 | 2.0 | 144 | 1.051 | 0.137 | 141 |
| Palladium chloride (1,000 ppm stock solution) $PdCl_2$ | — | — | — | — | — | 0.4 | — |
| Tetrabutylammonium bromide $C_{16}H_{36}BrN$ | 1 | 322 | 1.0 | 322 | — | — | 103 |
| Potassium carbonate $K_2CO_3$ | 3.7 | 138 | 3.7 | 511 | — | — | — |
| Water $H_2O$ | — | — | — | — | — | 1.6 | 100 |

### PERFORM THE REACTION

- ❑ In a 10-mL glass microwave reaction vessel place a magnetic stir bar, the tetrabutylammonium bromide, the 4-iodoacetophenone, and the potassium carbonate.
- ❑ Using a graduated cylinder add the water to the reaction vessel.
- ❑ Using an automatic delivery pipette, add the acrylic acid and palladium stock solution to the reaction vessel.

- ❑ Seal the reaction vessel with a cap according to the microwave manufacturer's recommendations.
- ❑ Shake the contents of the vessel to mix the reagents.
- ❑ Program the microwave unit to heat the vessel contents to 175°C using an initial microwave power of 250 W and hold at this temperature for 20 minutes.
- ❑ After the heating step is completed, allow the contents of the reaction vessel to cool to 50°C or below before removing it from the microwave cavity.

## Isolate the Product

- ❑ Add water (4 mL) to a clean 25-mL Erlenmeyer flask.
- ❑ Carefully open the reaction vessel and pipette the contents of the reaction vessel into the Erlenmeyer flask containing the water.
- ❑ Add 6 M HCl dropwise to the Erlenmeyer flask until the pH reaches 6. The product will precipitate out of the solution.
- ❑ Cool the solution in the ice bath for 10 minutes after the addition is complete.
- ❑ While the solution is cooling, set up a vacuum-filtration apparatus with a Hirsch funnel, side-arm flask, rubber collar, and a length of rubber vacuum tubing.
- ❑ Cool some distilled water (10 mL) in a 25-mL Erlenmeyer flask by placing it in an ice bath for washing the product after filtration.
- ❑ Connect the filtration system to a vacuum and place the correct size filter paper in the funnel.
- ❑ Wet the filter paper with a few drops of the cold distilled water and start the vacuum to seal the filter paper in place.
- ❑ Filter the reaction mixture by pouring the contents of the 25-mL Erlenmeyer flask into the funnel; transfer as much solid as possible.
- ❑ Rinse the flask with cold distilled water (3 mL) and add it to the filter funnel.
- ❑ Rinse the precipitate with additional cold distilled water (3 mL).
- ❑ Allow the precipitate to dry on the funnel for 10 minutes.
- ❑ Transfer the solid to a large piece of filter paper to dry completely.
- ❑ Weigh the dry product and calculate the yield and percent yield.

## Characterize the Product

- ❑ Determine the melting point of the white product.
- ❑ If the melting point indicates the product is not pure, re-crystallize the crude product from ethanol.
- ❑ Obtain an IR spectrum of the product and compare it with that of the starting material.
- ❑ Obtain $^{1}$H-NMR and $^{13}$C-NMR spectra in $d_6$-acetone if instructed to do so.

## PROCEDURE FOR USE IN A MULTIMODE MICROWAVE UNIT

### SYNTHESIS OF 4-ACETYLCINNAMIC ACID (2*E*-3-(4-ACETYLPHENYL)-2-PROPENOIC ACID)

[Pd]

$K_2CO_3$, $H_2O$

**Caution:** Acrylic acid is lachrymatory and an irritant. It should be measured out in a fume hood or well-ventilated area. 4-Iodoacetophenone is an irritant. The palladium stock solution is toxic. The use of goggles with side-shields, lab coats, and gloves is considered minimum and nondiscretionary safety practice in the laboratory.

### Table of Reagents and Physical Constants

| Reagent | Equiv. | FW | mmol | Mass (mg) | Density (g/mL) | Vol. (mL) | Mp/bp°C |
|---|---|---|---|---|---|---|---|
| 4-Iodoacetophenone $C_8H_7IO$ | 1 | 246 | 1.0 | 246 | — | — | 82–84 |
| Acrylic acid $C_3H_4O_2$ | 2 | 72.0 | 2.0 | 144 | 1.051 | 0.137 | 141 |
| Palladium chloride (1,000 ppm stock solution) $PdCl_2$ | — | — | — | — | — | 0.4 | — |
| Tetrabutylammonium bromide $C_{16}H_{36}BrN$ | 1 | 322 | 1.0 | 322 | — | — | 103 |
| Potassium carbonate $K_2CO_3$ | 3.7 | 138 | 3.7 | 511 | — | — | — |
| Water $H_2O$ | — | — | — | — | — | 4.0 | 100 |

### PERFORM THE REACTION

- ❑ In a 25-mL glass microwave reaction vessel place a magnetic stir bar, the tetrabutylammonium bromide, the 4-iodoanisole, and the potassium carbonate.
- ❑ Using a graduated cylinder add the water to the reaction vessel.
- ❑ Using an automatic delivery pipette, add the acrylic acid and palladium stock solution to the reaction vessel.
- ❑ Seal the reaction vessel with a cap according to the microwave manufacturer's recommendations.

- ❑ Place the sealed reaction vessel in the carousel, noting the vessel's position number and ensuring that vessels are evenly spaced around the carousel.
- ❑ When all the group's reaction vessels are in place, load the carousel into the microwave cavity.
- ❑ If provided by the manufacturer, connect the temperature probe to the control vessel.
- ❑ Program the microwave unit to heat the vessel contents to 175°C using an initial microwave power of 600 W and hold at this temperature for 20 minutes.
- ❑ After the heating step is completed, allow the contents of the reaction vessel to cool to 50°C or below before removing it from the microwave cavity.

### Isolate the Product

- ❑ Add water (4 mL) to a clean 25-mL Erlenmeyer flask.
- ❑ Carefully open the reaction vessel and pipette the contents of the reaction vessel into the Erlenmeyer flask containing the water.
- ❑ Add 6 M HCl dropwise to the Erlenmeyer flask until the pH reaches 6. The product will precipitate out of the solution.
- ❑ Cool the solution in the ice bath for 10 minutes after the addition is complete.
- ❑ While the solution is cooling, set up a vacuum-filtration apparatus with a Hirsch funnel, side-arm flask, rubber collar, and a length of rubber vacuum tubing.
- ❑ Cool some distilled water (10 mL) in a 25-mL Erlenmeyer flask by placing it in an ice bath for washing the product after filtration.
- ❑ Connect the filtration system to a vacuum and place the correct size filter paper in the funnel.
- ❑ Wet the filter paper with a few drops of the cold distilled water and start the vacuum to seal the filter paper in place.
- ❑ Filter the reaction mixture by pouring the contents of the 25-mL Erlenmeyer flask into the funnel; transfer as much solid as possible.
- ❑ Rinse the flask with cold distilled water (3 mL) and add it to the filter funnel.
- ❑ Rinse the precipitate with additional cold distilled water (3 mL).
- ❑ Allow the precipitate to dry on the funnel for 10 minutes.
- ❑ Transfer the solid to a large piece of filter paper to dry completely.
- ❑ Weigh the dry product and calculate the yield and percent yield.

### Characterize the Product

- ❑ Determine the melting point of the white product.
- ❑ If the melting point indicates the product is not pure, re-crystallize the crude product from ethanol.
- ❑ Obtain an IR spectrum of the product and compare it with that of the starting material.
- ❑ Obtain $^{1}$H-NMR and $^{13}$C-NMR spectra in $d_6$-acetone if instructed to do so.

## QUESTIONS

1. The product of the Heck reaction is an alkene. One of the starting materials is also an alkene. Why is there no competition between the product alkene and the starting material alkene in reacting with the aryl halide?
2. Predict the product of the Heck coupling of cyclopentene and bromobenzene.
3. Give two set of starting materials that could be used to prepare *trans*-4-methoxystilbene using a Heck reaction.
4. Draw a curved-arrow mechanism for the base-catalyzed hydrolysis of the methyl ester to form the carboxylic acid in the preparation of 4-methoxycinnamic acid.

# 14 Preparation of an Aryl Nitrile

## *Application of a Copper-Catalyzed Cyanation Reaction*[*]

**LEARNING GOALS**

- To synthesize an aryl nitrile compound
- To use a "green chemistry" approach in reactions involving cyanide
- To use extraction techniques in the isolation of a product
- To use thin-layer chromatography to determine product purity

## INTRODUCTION

Aryl nitriles find applications as dyes, herbicides, agrochemicals, and pharmaceuticals. In addition, they are also extremely useful intermediates for organic synthesis, mainly for the formation of heterocycles. Aryl nitriles can be prepared from aryl halides in a reaction called "cyanation". The chemistry dates back to the 1900s and first involved using copper(I) cyanide and heating at 260°C for 6 h, in a process called the Rosenmund–von Braun reaction. More recent modifications of the reaction allow for the use of copper or palladium catalysts but require sodium cyanide or potassium cyanide as reagents. The major drawback of all these procedures is the need for metal cyanide salts. They are very toxic, immediate medical attention being required in case of ingestion, skin absorption through open wounds, or inhalation of dust. In addition, alkali metal cyanides, in particular, can liberate hydrogen cyanide (HCN) upon reaction with acids. These safety concerns mean that handling and all reactions involving the preparation and use of metal and organic cyanide compounds should be carried out only by well-trained personnel in a fume hood, in full compliance with all local safety regulations.

[*] Experimental procedure developed by Catherine DeBlase and Michael Luciano, University of Connecticut and Merrimack College, respectively.

hal + CuCN —heat→ C≡N

R R

*aryl halide* *aryl nitrile*

*Rosenmund–von Braun reaction*

The discovery in 2004 that potassium hexacyanoferrate(II), $K_4[Fe(CN)_6]$, can be used as a cyanide source for making aryl nitriles from aryl halides represented a fundamental breakthrough. $K_4[Fe(CN)_6]$ has greatly reduced toxicity as compared to alkali metal cyanides. Indeed, the median lethal dose ($LD_{50}$) of $K_4[Fe(CN)_6]$ is lower than that for table salt. It does not liberate HCN, even when dissolved in dilute hydrochloric acid and boiled. However, under the conditions typically used in cyanation reactions it will liberate $CN^-$ very slowly and is therefore an effective cyanating reagent. All six of the $CN^-$ groups of $K_4[Fe(CN)_6]$ can be used for cyanation of aryl halides, harnessing the full potential of the reagent. Most procedures involve the use of palladium or copper complexes as catalysts for the reaction.

hal + [Fe(CN)₆]⁴⁻ 4 K⁺ —*Pd or Cu catalyst*, *heat*→ C≡N

R R

*potassium hexacyanoferrate (II)*

Cyanation reactions can be performed using water as the solvent. However, it is necessary to add either a phase-transfer catalyst or else a co-solvent to increase the solubility of the organic starting materials in the reaction mixture. One such co-solvent is tetraethylene glycol. A glycol is a compound in which two hydroxyl (OH) groups are attached to different carbon atoms. The smaller members of the glycol family are soluble in water but, because they contain alkyl chains too, they help dissolve organic compounds.

HO–O–O–O–OH *tetraethylene glycol*

## PROCEDURE FOR USE IN A MONOMODE MICROWAVE UNIT

### PREPARATION OF 1-CYANONAPHTHALENE

$$\text{1-iodonaphthalene} + K_4Fe(CN)_6 \xrightarrow[CuI]{MW} \text{1-cyanonaphthalene}$$

**Caution:** 1-Iodonapthalene and copper iodide are irritants. The use of goggles with side-shields, lab coats, and gloves is considered minimum and nondiscretionary safety practice in the laboratory.

### Table of Reagents and Physical Constants

| Reagent | Equiv. | FW | mmol | Mass (mg) | Density (g/mL) | Vol. (mL) | mp/bp°C |
|---|---|---|---|---|---|---|---|
| 1-Iodonaphthalene $C_{10}H_7I$ | 1 | 254 | 1.00 | 254 | — | — | 163 |
| Copper(I) iodide CuI | 0.15 | 190 | 0.15 | 28 | 5.62 | — | 605 |
| Potassium hexacyanoferrate(II) $K_4Fe(CN)_6$ | 0.3 | 422 | 0.30 | 127 | — | — | 70 |
| Tetraethylene glycol $C_8H_{18}O_5$ | — | — | — | — | — | 1.0 | 314 |
| Water $H_2O$ | — | — | — | — | — | 1.0 | 100 |

### PERFORM THE REACTION

- ❑ In a 10-mL glass microwave reaction vessel place a magnetic stir bar, the 1-iodonaphthalene, potassium hexacyanoferrate(II), and copper(I) iodide.
- ❑ Add the tetraethylene glycol and water to the reaction vessel using a graduated cylinder.
- ❑ Seal the reaction vessel with a cap according to the microwave manufacturer's recommendations.
- ❑ Place the sealed reaction vessel in the microwave cavity.
- ❑ Program the microwave unit to heat the vessel contents to 175°C using an initial microwave power of 150 W and hold at this temperature for 30 minutes.
- ❑ After the heating step is completed, allow the contents of the reaction vessel to cool to 50°C or below before removing it from the microwave cavity.

## Isolate the Product

- ❑ While the reaction is cooling, obtain two 25-mL Erlenmeyer flasks and label them "aqueous phase" and "organic phase".
- ❑ Carefully open the reaction vessel.
- ❑ Use a Pasteur pipette to transfer the reaction mixture to a 60-mL separatory funnel, leaving the magnetic stir bar in the reaction vessel.
- ❑ Rinse the reaction vessel with acetonitrile (3 mL).
- ❑ Pipette the rinse into the separatory funnel that contains the reaction mixture.
- ❑ Repeat the rinse of the reaction vessel with acetonitrile (3 mL) and add it to the separatory funnel as before.
- ❑ Rinse the reaction vessel with diethyl ether (3 mL).
- ❑ Pipette the rinse into the separatory funnel that contains the reaction mixture.
- ❑ Repeat the rinse of the reaction vessel with diethyl ether (3 mL) and add it to the separatory funnel as before.
- ❑ Rinse the reaction vessel with water (3 mL).
- ❑ Pipette the rinse into the separatory funnel that contains the reaction mixture.
- ❑ Add to the separatory funnel, additional diethyl ether (3 mL) and water (6 mL).
- ❑ Carefully stopper the separatory funnel and invert the funnel.
- ❑ Immediately vent the funnel by opening the stopcock to release pressure that may have developed.
- ❑ Close the stopcock and mix the two layers several times by inverting the funnel repeatedly.
- ❑ Vent the funnel as before.
- ❑ Close the stopcock and re-clamp the separatory funnel to the ring stand and remove the stopper.
- ❑ Allow the layers to separate.
- ❑ Move the bottom aqueous layer through the stopcock into the 25-mL Erlenmeyer flask labeled "aqueous phase".
- ❑ Add saturated sodium chloride solution (10 mL) to the organic phase in the separatory funnel.
- ❑ Stopper the funnel and mix the two phases by inverting the separatory funnel venting the funnel to release any pressure as before.
- ❑ Re-clamp the separatory funnel and remove the stopper.
- ❑ Move the bottom aqueous phase through the stopcock into the flask labeled "aqueous phase".
- ❑ Pour the top organic layer from the separatory funnel into the 25-mL Erlenmeyer flask labeled "organic phase".
- ❑ Add anhydrous magnesium sulfate ($MgSO_4$) to the flask containing the organic layer and stopper the flask.
- ❑ Allow the flask to sit for 10 minutes to complete the drying process.
- ❑ Tare a 50-mL round-bottom flask while the organic layer is drying.
- ❑ Transfer the organic layer to the round-bottom flask with a filter pipette; avoid transferring any of the drying agent.

- ❑ Rinse the Erlenmeyer flask and drying agent with diethyl ether (3 mL) and add this solution to the solution in the round-bottom flask.
- ❑ To determine the purity of the product, run a TLC of this solution by removing a small sample from the round-bottom flask with a micropipette.
- ❑ Spot the solution on a prepared TLC plate.
- ❑ Prepare a TLC chamber by taking a beaker large enough to fit the TLC plate and adding a small layer of hexanes.
- ❑ Cover the beaker with a watch glass.
- ❑ Carefully place the TLC plate in the beaker leaning the plate against the side of the beaker.
- ❑ Be sure the spot is above the level of the eluting solvent.
- ❑ Re-cover the beaker.
- ❑ Once the solvent front is near the top of the plate carefully remove the plate from the chamber.
- ❑ Immediately mark the solvent front with a pencil and then allow the plate to dry.
- ❑ Visualize the TLC plate with a UV lamp and mark any spots observed.
- ❑ Remove the diethyl ether from the organic solution under reduced pressure until constant weight is observed.
- ❑ Re-weigh the flask containing the product; calculate the yield and percent yield.

## Characterize the Product

- ❑ Obtain an IR spectrum of the product and compare it with that of the starting material.
- ❑ Obtain $^{1}$H-NMR and $^{13}$C-NMR spectra in $CDCl_3$ if instructed to do so.

## PROCEDURE FOR USE IN A MULTIMODE MICROWAVE UNIT

### PREPARATION OF 1-CYANONAPHTHALENE

1-Iodonaphthalene + $K_4Fe(CN)_6$ —(MW, CuI)→ 1-Cyanonaphthalene (CN)

**Caution:** 1-Iodonapthalene and copper iodide are irritants. The use of goggles with side-shields, lab coats, and gloves is considered minimum and nondiscretionary safety practice in the laboratory.

### Table of Reagents and Physical Constants

| Reagent | Equiv. | FW | mmol | Mass (mg) | Density (g/mL) | Vol. (mL) | mp/bp°C |
|---|---|---|---|---|---|---|---|
| 1-Iodonaphthalene $C_{10}H_7I$ | 1 | 254 | 1.00 | 254 | — | — | 163 |
| Copper(I) iodide CuI | 0.15 | 190 | 0.15 | 28 | 5.62 | — | 605 |
| Potassium hexacyanoferrate(II) $K_4Fe(CN)_6$ | 0.3 | 422 | 0.30 | 127 | — | — | 70 |
| Tetraethylene glycol $C_8H_{18}O_5$ | — | — | — | — | — | 2.0 | 314 |
| Water $H_2O$ | — | — | — | — | — | 2.0 | 100 |

### PERFORM THE REACTION

- ❑ In a 25-mL glass microwave reaction vessel place a magnetic stir bar, the 1-iodonaphthalene, potassium hexacyanoferrate(II), and copper(I) iodide.
- ❑ Add the tetraethylene glycol and water to the reaction vessel using a graduated cylinder.
- ❑ Seal the reaction vessel with a cap according to the microwave manufacturer's recommendations.
- ❑ Place the sealed reaction vessel in the carousel, noting the vessel's position number and ensuring that vessels are evenly spaced around the carousel.
- ❑ When all the group's reaction vessels are in place, load the carousel into the microwave cavity.
- ❑ If provided by the manufacturer, connect the temperature probe to the control vessel.
- ❑ Program the microwave unit to heat the vessel contents to 175°C using an initial microwave power of 600 W and hold at this temperature for 30 minutes.

- ❑ After the heating step is completed, allow the contents of the reaction vessel to cool to 50°C or below before removing it from the microwave cavity.

## Isolate the Product

- ❑ While the reaction is cooling, obtain two 50-mL Erlenmeyer flasks and label them “aqueous phase” and “organic phase”.
- ❑ Carefully open the reaction vessel.
- ❑ Use a Pasteur pipette to transfer the reaction mixture to a 125-mL separatory funnel, leaving the magnetic stir bar in the reaction vessel.
- ❑ Rinse the reaction vessel with acetonitrile (5 mL).
- ❑ Pipette the rinse into the separatory funnel that contains the reaction mixture.
- ❑ Repeat the rinse of the reaction vessel with acetonitrile (5 mL) and add it to the separatory funnel as before.
- ❑ Rinse the reaction vessel with diethyl ether (5 mL).
- ❑ Pipette the rinse into the separatory funnel that contains the reaction mixture.
- ❑ Repeat the rinse of the reaction vessel with diethyl ether (5 mL) and add it to the separatory funnel as before.
- ❑ Rinse the reaction vessel with water (5 mL).
- ❑ Pipette the rinse into the separatory funnel that contains the reaction mixture.
- ❑ Add to the separatory funnel, additional diethyl ether (5 mL) and water (10 mL).
- ❑ Carefully stopper the separatory funnel and invert the funnel.
- ❑ Immediately vent the funnel by opening the stopcock to release pressure that may have developed.
- ❑ Close the stopcock and mix the two layers several times by inverting the funnel repeatedly.
- ❑ Vent the funnel as before.
- ❑ Close the stopcock and re-clamp the separatory funnel to the ring stand and remove the stopper.
- ❑ Allow the layers to separate.
- ❑ Move the bottom aqueous layer through the stopcock into the 50-mL Erlenmeyer flask labeled “aqueous phase”.
- ❑ Add saturated sodium chloride solution (10 mL) to the organic phase in the separatory funnel.
- ❑ Stopper the funnel and mix the two phases by inverting the separatory funnel venting the funnel to release any pressure as before.
- ❑ Re-clamp the separatory funnel and remove the stopper.
- ❑ Move the bottom aqueous phase through the stopcock into the flask labeled “aqueous phase”.
- ❑ Pour the top organic layer from the separatory funnel into the 50-mL Erlenmeyer flask labeled “organic phase”.

- ❑ Add anhydrous magnesium sulfate ($MgSO_4$) to the flask containing the organic layer and stopper the flask.
- ❑ Allow the flask to sit for 10 minutes to complete the drying process.
- ❑ Tare a 50-mL round-bottom flask while the organic layer is drying.
- ❑ Transfer the organic layer to the round-bottom flask with a filter pipette; avoid transferring any of the drying agent.
- ❑ Rinse the Erlenmeyer flask and drying agent with diethyl ether (3 mL) and add this solution to the solution in the round-bottom flask.
- ❑ To determine the purity of the product, run a TLC of this solution by removing a small sample from the round-bottom flask with a micropipette.
- ❑ Spot the solution on a prepared TLC plate.
- ❑ Prepare a TLC chamber by taking a beaker large enough to fit the TLC plate and adding a small layer of hexanes.
- ❑ Cover the beaker with a watch glass.
- ❑ Carefully place the TLC plate in the beaker leaning the plate against the side of the beaker.
- ❑ Be sure the spot is above the level of the eluting solvent.
- ❑ Re-cover the beaker.
- ❑ Once the solvent front is near the top of the plate carefully remove the plate from the chamber.
- ❑ Immediately mark the solvent front with a pencil and then allow the plate to dry.
- ❑ Visualize the TLC plate with a UV lamp and mark any spots observed.
- ❑ Remove the diethyl ether from the organic solution under reduced pressure until constant weight is observed.
- ❑ Re-weigh the flask containing the product; calculate the yield and percent yield.

## Characterize the Product

- ❑ Obtain an IR spectrum of the product and compare it with that of the starting material.
- ❑ Obtain $^{1}H$-NMR and $^{13}C$-NMR spectra in $CDCl_3$ if instructed to do so.

## QUESTIONS

1. Nitriles are classed as carboxylic acid derivatives. However, they do not contain a carbonyl group, unlike esters, amides, acid chlorides, and acid anhydrides. Explain why nitriles are still classed in this group of carboxylic acid derivatives.
2. Aryl nitriles are useful starting materials for the formation of heterocycles. Find an example.
3. Why are metal cyanide salts such as potassium cyanide so toxic to humans?
4. The reaction of 1-iodonapthalene with $K_4[Fe(CN)_6]$ requires the use of a catalyst. What is the definition of a "catalyst" and what does it do to the reaction?

# 15 Alkene Metathesis

## *Preparation of a Substituted Cyclopentene**

**LEARNING GOALS**

- To prepare a cyclic compound
- To perform a metal-catalyzed reaction
- To isolate a product by means of microscale flash column chromatography

## INTRODUCTION

The use of metal catalysts in organic chemistry has dramatically increased the number of transformations that can be performed. In 2005, three chemists were awarded the Nobel Prize in Chemistry for their pioneering work in this field, specifically the development of metal-catalyzed alkene metathesis. The word metathesis essentially means to change places and has been compared to a dance in which couples change partners. In alkene metathesis reactions, double bonds are broken and formed between carbon atoms in ways that cause substituents to change places.

$$(R^1)_2C{=}C(R^2)_2 + (R^3)_2C{=}C(R^4)_2 \xrightarrow{\textit{catalyst}} (R^1)_2C{=}C(R^4)_2 + (R^3)_2C{=}C(R^2)_2$$

Alkene metathesis can be used in a wide range of different applications. It can be used to make new alkenes (self metathesis and cross metathesis), to make ring compounds (ring-closing metathesis), or to open ring compounds (ring-opening metathesis). A variant of ring-opening metathesis can also be used to make polymers (ring-opening metathesis polymerization).

* Modified from a conventional procedure developed by Michael Mercadante and Christopher Kelly, University of Connecticut.

self metathesis

catalyst

cross methathesis

catalyst

ring-opening methathesis (ROM)

catalyst

ring-opening methathesis (ROM)

catalyst

catalyst

ring-opening methathesis polymetization (ROM)

A problem that can be encountered in alkene metathesis reactions is that a distribution of products is obtained. This is because both the forward and backward reactions are possible. As with most chemical reactions, alkene metathesis is driven by thermodynamics. The distribution of products is determined by the relative energies of the possible products. However, all the possible products may have similar energy values because all of them contain an alkene. A way to drive the reaction to products and to completion is for one of the products to be a gas. Thus, many alkene metathesis reactions often involve α-alkenes, that is, those bearing two hydrogen atoms attached to one end of the double bond. That way, one of the products of alkene metathesis is ethene, a gas. Cross-metathesis and ring-closing metathesis benefit in this regard.

catalyst

A wide range of metal catalysts has been developed for metathesis reactions. They can be broken down into two main classes: those that are molybdenum-based and those that are ruthenium-based. A common theme to all the catalysts is that they are metal alkylidenes, containing a metal-to-carbon double bond.

labile ligand

metal alkylidene functionality

Mo-based catalyst (Schrock catalyst)

Ru-based catalyst (Grubbs catalyst)

The catalysts also contain at least one labile ligand, a group that readily dissociates from the central metal atom to allow the alkene substrate to bind. This is key to the mechanism of the reaction that involves the formation of a metallocyclobutane, an analog of a cyclobutane but with a metal in place of one of the carbon atoms in the four-membered ring.

***initiation***

M=C R R L + R¹ ⇌ M=C R R R¹ ⇌ M C R R R¹ C C H H H → M=C R¹ H + R R

*metallocyclobutane*

***catalytic cycle***

M C R¹ H H C C R¹ H H

H R¹ R¹ H

H R¹ H H

M=C H H + H R¹ H H

M C H H R¹ C C H H H

H R¹ H H

H H H H

M=C R¹ H + H R¹ H H

## PROCEDURE FOR USE IN A MONOMODE MICROWAVE UNIT

### Preparation of Diethyl Cyclopent-3-Ene-1,1-Dicarboxylate

MW

*Grubbs–Hoveyda catalyst (second generation)*

**Caution:** Diethyl diallylmalonate is an irritant and dichloromethane is a flammable organic solvent. Transition-metal complexes such as the second-generation Grubbs–Hoveyda catalyst should be handled with care due to their possible toxicity. The use of goggles with side-shields, lab coats, and gloves is considered minimum and nondiscretionary safety practice in the laboratory.

**Table of Reagents and Physical Constants**

| Reagent | Equiv. | FW | mmol | Mass (mg) | Density (g/mL) | Vol. (mL) | mp/bp°C |
|---|---|---|---|---|---|---|---|
| Diethyl diallylmalonate $C_{13}H_{20}O_4$ | 1 | 240 | 0.633 | 152 | 0.99 | 0.15 | 128 |
| Grubbs–Hoveyda catalyst (second generation) $C_{31}H_{38}Cl_2N_2ORu$ | — | 627 | 0.019 | 12 | — | — | 216–220 |
| Dichloromethane $CH_2Cl_2$ | — | — | — | — | — | 4 | 40 |

### Perform the Reaction

- ❑ In a 10-mL glass microwave reaction vessel, place a magnetic stir bar, the Grubbs–Hoveyda catalyst, and the dichloromethane.
- ❑ Using an automatic delivery pipette add the diethyl diallylmalonate.
- ❑ Seal the reaction vessel with a cap according to the microwave manufacturer's recommendations.
- ❑ Place the sealed reaction vessel in the microwave cavity.
- ❑ Program the microwave unit to heat the vessel contents to 65°C over a 3-minute ramp period and then hold at this temperature for 25 minutes.
- ❑ After the heating step is completed, allow the contents of the reaction vessel to cool to 50°C or below before removing it from the microwave cavity.

## Isolate the Product

- ❑ Carefully open the reaction vessel.
- ❑ Pour the contents of the reaction vessel into a 25-mL round-bottom flask.
- ❑ Remove the dichloromethane solvent either by rotary evaporation under reduced pressure or by placing the flask and its contents in a steam bath in a fume hood.
- ❑ Plug a 5.75-mm disposable glass pipette with a small piece of glass wool.
- ❑ Add to the pipette, silica gel to a height of 4 cm.
- ❑ Clamp the pipette so that it sits just above a clean-tared round-bottom or pear-shaped flask.
- ❑ Dissolve the remaining contents of the flask in the minimum volume of hexane/ethyl acetate (8:2).
- ❑ Load the hexane/ethyl acetate solution onto the silica gel in the pipette.
- ❑ Elute the product through the silica by using 7:3 hexane/ethyl acetate (10 mL), collecting the solution in the tared flask.
- ❑ Remove the solvent under reduced pressure until a constant weight is observed.
- ❑ Re-weigh the flask containing the light-yellow oil.
- ❑ Calculate the yield and percent yield.

## Characterize the Product

- ❑ Obtain $^{1}H$- and $^{13}C$-NMR spectra in $CDCl_3$.
- ❑ Determine the purity of the product by GC-MS if instructed to do so.

## PROCEDURE FOR USE IN A MULTIMODE MICROWAVE UNIT

### PREPARATION OF DIETHYL CYCLOPENT-3-ENE-1,1-DICARBOXYLATE

*Grubbs–Hoveyda catalyst (second generation)*

**Caution:** Diethyl diallylmalonate is an irritant and dichloromethane is a flammable organic solvent. Transition-metal complexes such as the second-generation Grubbs–Hoveyda catalyst should be handled with care due to their possible toxicity. The use of goggles with side-shields, lab coats, and gloves is considered minimum and nondiscretionary safety practice in the laboratory.

### Table of Reagents and Physical Constants

| Reagent | Equiv. | FW | mmol | Mass (mg) | Density (g/mL) | Vol. (mL) | mp/bp°C |
|---|---|---|---|---|---|---|---|
| Diethyl diallylmalonate $C_{13}H_{20}O_4$ | 1 | 240 | 0.633 | 152 | 0.99 | 0.15 | 128 |
| Grubbs–Hoveyda catalyst (second generation) $C_{31}H_{38}Cl_2N_2ORu$ | — | 627 | 0.019 | 12 | — | — | 216–220 |
| Dichloromethane $CH_2Cl_2$ | — | — | — | — | — | 4 | 40 |

### PERFORM THE REACTION

- ❑ In a 25-mL glass microwave reaction vessel, place a magnetic stir bar, the Grubbs–Hoveyda catalyst, and the dichloromethane.
- ❑ Using an automatic delivery pipette add the diethyl diallylmalonate.
- ❑ Seal the reaction vessel with a cap according to the microwave manufacturer's recommendations.
- ❑ Place the sealed reaction vessel in the carousel, noting the vessel's position number and ensuring that vessels are evenly spaced around the carousel.
- ❑ When all the group's reaction vessels are in place, load the carousel into the microwave cavity.

- ❑ If provided by the manufacturer, connect the temperature probe to the control vessel.
- ❑ Program the microwave unit to heat the vessel contents to 65°C over a 5-minute ramp period and then hold at this temperature for 25 minutes.
- ❑ After the heating step is completed, allow the contents of the reaction vessel to cool to 50°C or below before removing it from the microwave cavity.

## Isolate the Product

- ❑ Carefully open the reaction vessel.
- ❑ Pour the contents of the reaction vessel into a 25-mL round-bottom flask.
- ❑ Remove the dichloromethane solvent either by rotary evaporation under reduced pressure or by placing the flask and its contents in a steam bath in a fume hood.
- ❑ Plug a 5.75-mm disposable glass pipette with a small piece of glass wool.
- ❑ Add to the pipette, silica gel to a height of 4 cm.
- ❑ Clamp the pipette so that it sits just above a clean-tared round-bottom or pear-shaped flask.
- ❑ Dissolve the remaining contents of the flask in the minimum volume of hexane/ethyl acetate (8:2).
- ❑ Load the hexane/ethyl acetate solution onto the silica gel in the pipette.
- ❑ Elute the product through the silica by using 7:3 hexane/ethyl acetate (10 mL), collecting the solution in the tared flask.
- ❑ Remove the solvent under reduced pressure until a constant weight is observed.
- ❑ Re-weigh the flask containing the light-yellow oil.
- ❑ Calculate the yield and percent yield.

## Characterize the Product

- ❑ Obtain $^{1}H$- and $^{13}C$-NMR spectra in $CDCl_3$.
- ❑ Determine the purity of the product by GC-MS if instructed to do so.

## QUESTIONS

1. What is the by-product of your reaction?
2. At what substrate concentration was the reaction performed? Why was it not performed at a higher concentration that is more typical for organic reactions?
3. By researching the literature, find three practical applications of alkene metathesis.
4. Assign the signals in your $^1$H-NMR to the proton environments on the product of the reaction.
5. Here is an example of an olefin metathesis reaction. Show how the product is formed from the starting material in this case.

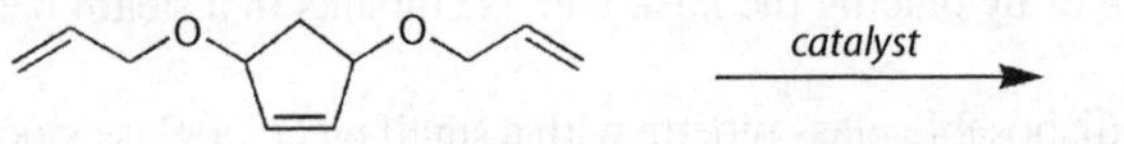

*catalyst*

# 16 Click Reaction

## *Preparation of a Triazole**

**LEARNING GOALS**

- To prepare a heterocyclic compound
- To perform a metal-catalyzed reaction
- To understand the concept of atom efficiency

## INTRODUCTION

In an ideal reaction, the moles of starting materials or reactants would equal those of all products generated and no atom would be wasted. As such, this reaction would be termed 100% atom efficient. The metric of atom efficiency (also known as atom economy) describes the conversion efficiency of a chemical process in terms of all atoms involved. It can be defined as

$$\text{atom efficiency} = 100 \times \frac{\textit{molecular weight of desired product}}{\textit{molecular weight of all starting materials}}$$

The atom economy of a reaction can be poor even if the chemical yield is 100%. Take, for example, the Wittig reaction: a significant portion of the starting materials do not end up in the desired product. Of the reactions used industrially, the catalytic hydrogenation of alkenes comes very close to being an ideal atom economic reaction.

* Developed by Neha Grewal, University of Connecticut. Modified from a conventional procedure: Sharpless, W. D.; Wu, P.; Hansen, T. V.; Lindberg, J. G. *J. Chem. Educ.*, 2005, 82, 1833–1836.

***Wittig reaction***

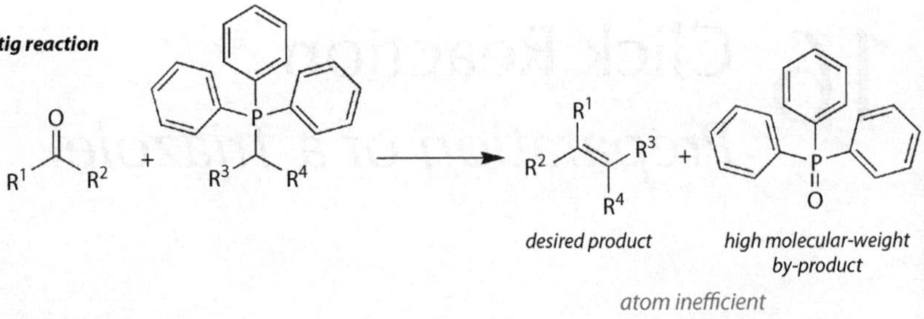

desired product

high molecular-weight by-product

atom inefficient

***catalytic hydrogenation***

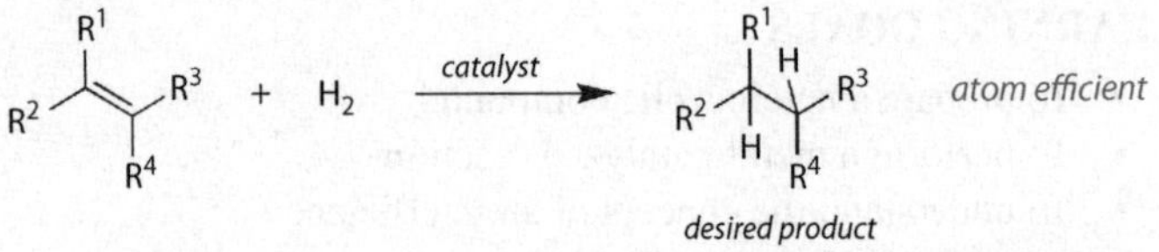

catalyst

desired product

atom efficient

Another class of atom-efficient reactions is the so-called "click" processes. "Click chemistry" is a term that was introduced by K. B. Sharpless in 2001 to describe reactions that are high yielding, wide in scope, create only by-products that can be removed without chromatography, are stereospecific, simple to perform, and can be conducted in easily removable or benign solvents. Many cycloaddition reactions fall into this category. A cycloaddition reaction involves two or more unsaturated molecules (or parts of the same molecule) combining with the formation of a cyclic adduct in which there is a net reduction of the bond multiplicity. One such example is the 1,3-dipolar cycloaddition of an organic azide and an alkyne to yield a 1,2,3-triazole. This reaction, known as the "Huisgen cycloaddition," requires elevated temperatures and long reaction times. It is called a 1,3-dipolar cycloaddition because azides are dipolar compounds with delocalized electrons and a separation of charge over three atoms.

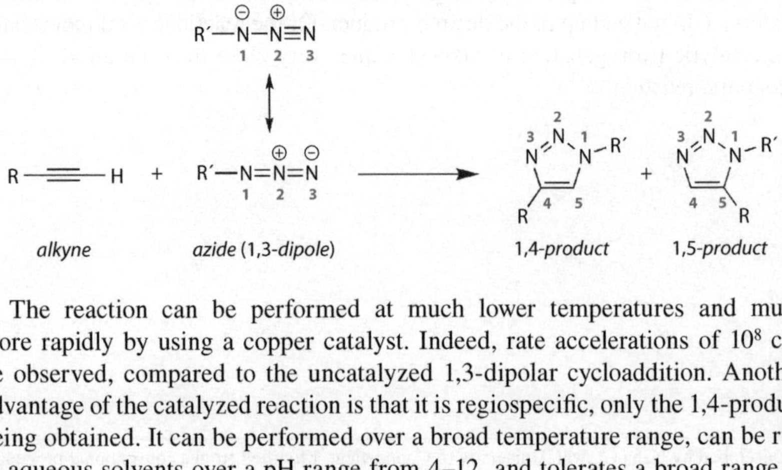

*alkyne* *azide* (1,3-*dipole*) 1,4-*product* 1,5-*product*

The reaction can be performed at much lower temperatures and much more rapidly by using a copper catalyst. Indeed, rate accelerations of $10^8$ can be observed, compared to the uncatalyzed 1,3-dipolar cycloaddition. Another advantage of the catalyzed reaction is that it is regiospecific, only the 1,4-product being obtained. It can be performed over a broad temperature range, can be run in aqueous solvents over a pH range from 4–12, and tolerates a broad range of functional groups. The product can be isolated by simple filtration or extraction without the need for chromatography or re-crystallization. The active copper(I) catalyst is generated in situ from a Cu(II) salt such as copper(II) sulfate using sodium ascorbate as a reducing agent. Organic azides are explosive and so are best generated in situ by reaction of an alkyl halide with sodium azide.

$$R{-}C{\equiv}C{-}H \;+\; R'{-}N{=}\overset{\oplus}{N}{=}\overset{\ominus}{N} \xrightarrow[\text{water}]{CuSO_4,\ \textit{sodium ascorbate}}$$

$$R'{-}Br \;+\; NaN_3 \xrightarrow{-NaBr} R'{-}N_3$$

## PROCEDURE FOR USE IN A MONOMODE MICROWAVE UNIT

### Preparation of 1-Phenyl-2-(4-Phenyl-[1,2,3]Triazol-1-yl)-Ethanone

+ + $NaN_3$ → MW; $CuSO_4$, Na ascorbate; water/t-butanol

**Caution:** Sodium azide is very hazardous. When mixed with acid, it rapidly reacts to form hydrazoic acid, a highly toxic gas with a pungent odor. The mixing of sodium azide with any acidic solution must be avoided at all times. Organic azides are potentially explosive. Any azide-containing waste solutions should be handled separately from other chemical wastes. Phenylacetylene is toxic and flammable. It should be dispensed in a hood. Phenacyl bromide (2-bromoacetophenone) is an irritant. The use of goggles with side-shields, lab coats, and gloves is considered minimum and nondiscretionary safety practice in the laboratory.

### Table of Reagents and Physical Constants

| Reagent | Equiv. | FW | mmol | Mass (mg) | Density (g/mL) | Vol. (mL) | mp/bp°C |
|---|---|---|---|---|---|---|---|
| 2-Bromoacetophenone $C_8H_7BrO$ | 1 | 199 | 2 | 398 | — | — | 48 |
| Phenylacetylene $C_8H_6$ | 1 | 102 | 2 | 204 | 0.93 | 0.20 | 142 |
| Sodium azide $NaN_3$ | 1.05 | 65.0 | 2.1 | 136 | — | — | 275 |
| Sodium ascorbate $C_6H_7NaO_6$ | 0.1 | 198 | 0.2 | 39 | — | — | 220 |
| Copper(II) sulfate solution (1M) $CuSO_4 \cdot 5H_2O$ | 0.05 | 240 | 0.1 | — | — | 0.10 | — |
| *t*-Butanol $C_4H_{10}O$ | — | — | — | — | — | 1.50 | 25/82 |
| Water $H_2O$ | — | — | — | — | — | 1.50 | 100 |

### Perform the Reaction

- ❑ In a 10-mL glass microwave reaction vessel, place a magnetic stir bar, the 2-bromoacetophenone, sodium azide, and sodium ascorbate.
- ❑ Using an automatic delivery pipette add the 1-M copper(II) sulfate solution to the reaction vessel.

- ❏ Using a graduated cylinder add the *tert*-butanol and water to the reaction vessel.
- ❏ Seal the reaction vessel with a cap according to the microwave manufacturer's recommendations.
- ❏ Place the sealed reaction vessel in the microwave cavity.
- ❏ Program the microwave unit to heat the vessel contents to 150°C over a 5-minute ramp period and then hold at this temperature for 20 minutes.
- ❏ After the heating step is completed, allow the contents of the reaction vessel to cool to 50°C or below before removing it from the microwave cavity.

## Isolate the Product

- ❏ While the solution is cooling set up a vacuum filtration system with a Hirsch funnel, side-arm flask, rubber collar, and a length of rubber vacuum tubing.
- ❏ In an ice bath, cool some water (4–8 mL) in a 25-mL Erlenmeyer flask for washing the precipitate after filtration.
- ❏ Place 10–20 g of crushed ice in a 50-mL beaker.
- ❏ Carefully open the reaction vessel.
- ❏ Pour the contents of the reaction vessel onto the crushed ice.
- ❏ Add ice water (1 mL) to the reaction vessel.
- ❏ Pour the washings into the beaker containing the crushed ice and reaction product.
- ❏ Remove the stirring bar from the beaker containing the crushed ice and reaction product with a magnetic retrieving wand or a pair of tweezers.
- ❏ Add a 10% ammonia solution (5 mL) to the beaker containing the crushed ice and reaction product.
- ❏ Connect the filtration system to a vacuum and place the correct size filter paper in the funnel.
- ❏ Wet the filter paper with a few drops of the cold water and start the vacuum to seal the filter paper in place.
- ❏ Filter the reaction mixture by pouring the contents of the beaker into the funnel; transfer as much solid as possible.
- ❏ Rinse the reaction vessel with cold water (2 mL) and add the washings to the filter funnel.
- ❏ Rinse the precipitate on the filter with additional cold water (1 mL).
- ❏ Allow the precipitate to dry on the funnel for 10 minutes.
- ❏ Transfer the solid to a large piece of filter paper to dry completely.
- ❏ Weigh the dry product and calculate the yield and percentage yield.

## Characterize the Product

- ❏ Determine the melting point of the product.
- ❏ Obtain a $^1$H-NMR spectrum in $d_6$-DMSO.

## PROCEDURE FOR USE IN A MULTIMODE MICROWAVE UNIT

### PREPARATION OF 1-PHENYL-2-(4-PHENYL-[1,2,3]TRIAZOL-1-YL)-ETHANONE



**Caution:** Sodium azide is very hazardous. When mixed with acid, it rapidly reacts to form hydrazoic acid, a highly toxic gas with a pungent odor. The mixing of sodium azide with any acidic solution must be avoided at all times. Organic azides are potentially explosive. Any azide-containing waste solutions should be handled separately from other chemical wastes. Phenylacetylene is toxic and flammable. It should be dispensed in a hood. Phenacyl bromide (2-bromoacetophenone) is an irritant. The use of goggles with side-shields, lab coats, and gloves is considered minimum and nondiscretionary safety practice in the laboratory.

### Table of Reagents and Physical Constants

| Reagent | Equiv. | FW | mmol | Mass (mg) | Density (g/mL) | Vol. (mL) | mp/bp°C |
|---|---|---|---|---|---|---|---|
| 2-Bromoacetophenone $C_8H_7BrO$ | 1 | 199 | 4 | 796 | — | — | 48 |
| Phenylacetylene $C_8H_6$ | 1 | 102 | 4 | 408 | 0.93 | 0.4 | 142 |
| Sodium azide $NaN_3$ | 1.05 | 65.0 | 4.2 | 272 | — | — | 275 |
| Sodium ascorbate $C_6H_7NaO_6$ | 0.1 | 198 | 0.4 | 78 | — | — | 220 |
| Copper(II) sulfate solution (1M) $CuSO_4 \cdot 5H_2O$ | 0.05 | 240 | 0.2 | — | — | 0.2 | — |
| *t*-Butanol $C_4H_{10}O$ | — | — | — | — | — | 3.0 | 25/82 |
| Water $H_2O$ | — | — | — | — | — | 3.0 | 100 |

### PERFORM THE REACTION

- ❑ In a 25-mL glass microwave reaction vessel, place a magnetic stir bar, the 2-bromoacetophenone, sodium azide, and sodium ascorbate.
- ❑ Using an automatic delivery pipette add the 1-M copper(II) sulfate solution to the reaction vessel.
- ❑ Using a graduated cylinder add the *t*-butanol and water to the reaction vessel.

- ❑ Seal the reaction vessel with a cap according to the microwave manufacturer's recommendations.
- ❑ Place the sealed reaction vessel in the carousel, noting the vessel's position number and ensuring that vessels are evenly spaced around the carousel.
- ❑ When all the group's reaction vessels are in place, load the carousel into the microwave cavity.
- ❑ If provided by the manufacturer, connect the temperature probe to the control vessel.
- ❑ Program the microwave unit to heat the vessel contents to 150°C over a 5-minute ramp period and then hold at this temperature for 20 minutes.
- ❑ After the heating step is completed, allow the contents of the reaction vessel to cool to 50°C or below before removing it from the microwave cavity.

### Isolate the Product

- ❑ While the solution is cooling set up a vacuum filtration system with a Hirsch funnel, side-arm flask, rubber collar, and a length of rubber vacuum tubing.
- ❑ In an ice bath, cool some water (10–20 mL) in a 50-mL Erlenmeyer flask for washing the precipitate after filtration.
- ❑ Place 20–30 g of crushed ice in a 100-mL beaker.
- ❑ Carefully open the reaction vessel.
- ❑ Pour the contents of the reaction vessel onto the crushed ice.
- ❑ Add ice water (2 mL) to the reaction vessel.
- ❑ Pour the washings into the beaker containing the crushed ice and reaction product.
- ❑ Remove the stirring bar from the beaker containing the crushed ice and reaction product with a magnetic retrieving wand or a pair of tweezers.
- ❑ Add 10% ammonia solution (10 mL) to the beaker containing the crushed ice and reaction product.
- ❑ Connect the filtration system to a vacuum and place the correct size filter paper in the funnel.
- ❑ Wet the filter paper with a few drops of the cold water and start the vacuum to seal the filter paper in place.
- ❑ Once the ice has melted, filter the reaction mixture by pouring the contents of the beaker into the funnel; transfer as much solid as possible.
- ❑ Rinse the reaction vessel with cold water (3 mL) and add the washings to the filter funnel.
- ❑ Rinse the precipitate on the filter with additional cold water (2 mL).
- ❑ Allow the precipitate to dry on the funnel for 10 minutes.
- ❑ Transfer the solid to a large piece of filter paper to dry completely.
- ❑ Weigh the dry product and calculate the yield and percentage yield.

### Characterize the Product

- ❑ Determine the melting point of the product.
- ❑ Obtain a $^{1}$H-NMR spectrum in $d_6$-DMSO.

## QUESTIONS

1. Justifying your answers, would you consider the following reactions atom efficient?
   (a) The addition of bromine to an alkene to yield a dibromoalkane
   (b) The palladium-catalyzed Suzuki coupling reaction
   (c) The Gabriel synthesis of alkyl amines from alkyl halides
   (d) The Diels–Alder reaction
2. By researching the literature, find three practical applications of the click reaction between alkynes and azides.
3. Assign the signals in your $^{1}$H-NMR to the proton environments on the product of the reaction.
4. Draw a mechanism for the uncatalyzed 1,3-dipolar cycloaddition (Huisgen) reaction between 1-propyne and methyl azide.

# 17 Coordination Chemistry
## *Preparation of Cisplatin**

**LEARNING GOALS**

- To prepare a coordination complex
- To prepare a pharmaceutical compound
- To understand the concepts of isomerism in transition metal complexes

## INTRODUCTION

Metals bind to a wide range of atoms and molecules, these being generically termed as ligands. Metal complexes bearing four ligands can adopt one of two possible geometries, namely square planar or tetrahedral. Square planar coordination is rare except in the case of metals with a d-electron configuration of $d^8$. This includes complexes of nickel, palladium, and platinum when the oxidation state of the metal is +2. When all the ligands around the central metal atom are the same, there is only one arrangement. However, when there are two different ligands attached to the metal and where there are two of each ligand, there are two possible isomeric structures. Looking at the complex MA2B2, where A and B are ligands attached to the central metal atom M, when both A ligands are next to each other the geometry is termed *cis*. When the two A ligands are opposite each other, the geometry is termed *trans*.

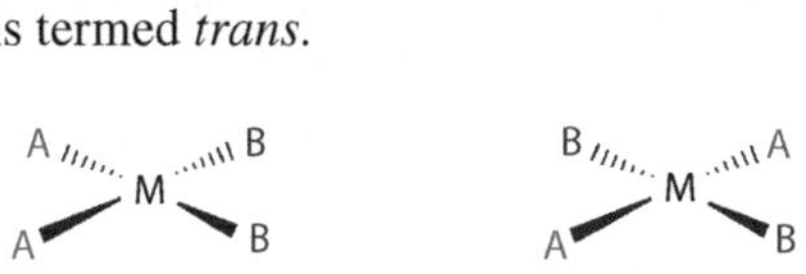

A, B, M, A, B — *cis isomer*

B, A, M, A, B — *trans isomer*

The objective of a chemotherapeutic drug is to kill cancer cells selectively, leaving healthy cells alone. This proves to be quite a challenge; a number of very complex drugs are being developed. In the 1960s Cisplatin, a very simple compound containing platinum, was serendipitously discovered to have a high activity for killing cancer cells and is still used, together with variants, today. A biologist and a biophysicist were studying the effects of electric fields on the growth of

* Procedure developed by Elizabeth Pedrick and Trevor Trueman, University of Connecticut.

bacteria. The bacteria around the platinum electrode used in their experiments seemed to be greatly elongated. They discovered that they had prevented cell division but not the other growth processes, which led to the elongation. By collaborating with chemists, researchers were able to deduce that this effect was due not to the electrical fields, but instead to a compound that was formed in a reaction between the platinum electrodes and components of the solution containing the bacteria. The compound was the simple square planar complex $PtCl_2(NH_3)_2$, bearing two chloride ligands and two ammonia ligands. However, only the *cis*-isomer of this complex is a potent chemotherapeutic drug. The *trans*-isomer is very much less active. Preparing $PtCl_2(NH_3)_2$ exclusively in the *cis*-geometry has been challenging, often requiring many steps. However, a one-step method has recently been developed. It uses potassium tetrachloroplatinate ($K_2PtCl_4$) as the platinum starting material, reacting it with potassium chloride and ammonium acetate.

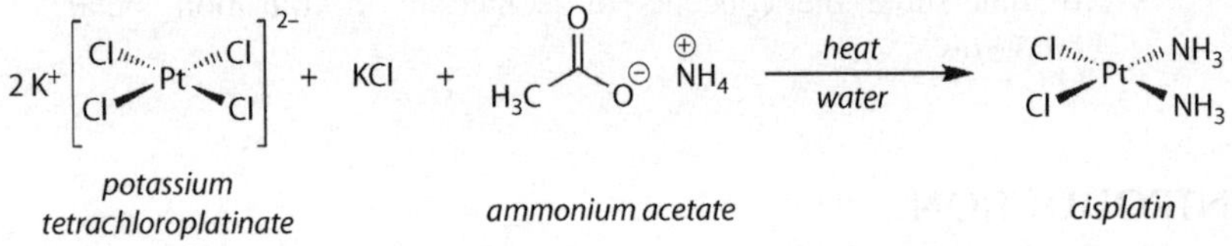

$$2\,K^+ [PtCl_4]^{2-} + KCl + H_3C\text{-}CO\text{-}O^- \; NH_4^+ \xrightarrow[\text{water}]{\text{heat}} cis\text{-}PtCl_2(NH_3)_2$$

*potassium tetrachloroplatinate* *ammonium acetate* *cisplatin*

In chemotherapy, cisplatin works by binding selectively to cancerous cells. All cells, be they healthy or cancerous, contain DNA. This DNA is made up of a series of repeating units called nucleotides and comprises two strands. The sequence of these units dictates the make-up of the cell. A key property of DNA is that it can replicate, or make copies of itself. Each strand of DNA can serve as a pattern for duplicating its sequence. This is critical when cells divide because each new cell needs to have an exact copy of the DNA present in the old cell. Cancerous cells often have a lot of places along the DNA strand where identical nucleotide units are next to one another, like words with double letters in them. Cisplatin binds strongly to these double letters and, by doing this, stops the cancer cells from reproducing. Inasmuch as healthy cells do not have the same double letters they are not affected as much. Cisplatin does, however, have some side effects associated with its use. To try to overcome these, as well as make it even more selective for cancer cells, a range of improved platinum-containing drugs has been developed more recently and is currently either in use or in clinical trials. Examples include carboplatin, oxaliplatin, and picoplatin.

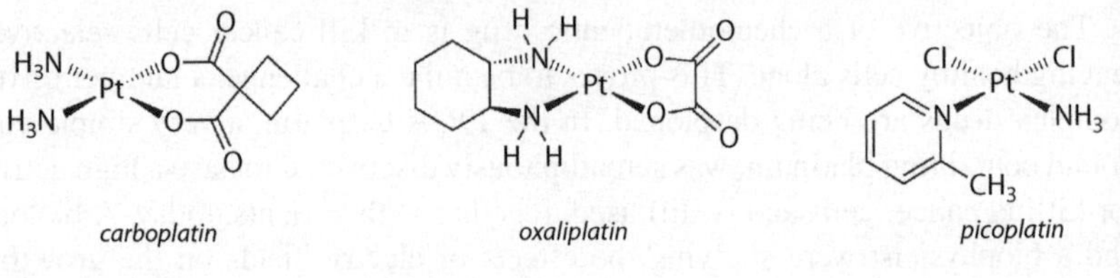

*carboplatin* *oxaliplatin* *picoplatin*

# PROCEDURE FOR USE IN A MONOMODE MICROWAVE UNIT

## Preparation of *cis*-Diamminedichloroplatinum(II) (Cisplatin)

$$K_2PtCl_4 + KCl + (NH_4)^+(CH_3CO_2)^- \xrightarrow{MW} cis\text{-}[PtCl_2(NH_3)_2]$$

**Caution:** Platinum salts are toxic and should be handled with care. The use of goggles with side-shields, lab coats, and gloves is considered minimum and non-discretionary safety practice in the laboratory.

### Table of Reagents and Physical Constants

| Reagent | Equiv. | FW | mmol | Mass (mg) | Vol. (mL) | mp/bp°C |
|---|---|---|---|---|---|---|
| Potassium tetrachloroplatinate(II) $K_2PtCl_4$ | 1 | 415 | 0.048 | 20 | — | 265 |
| Potassium chloride KCl | 5.6 | 74.5 | 0.27 | 20 | — | 770 |
| Ammonium acetate $C_2H_7NO_2$ | 4.3 | 77 | 0.21 | 16 | — | 114 |
| Water $H_2O$ | — | — | — | — | 1.0 | 100 |

## Perform the Reaction

- ❑ In a 10-mL glass microwave reaction vessel, place a magnetic stir bar, the potassium tetrachloroplatinate, potassium chloride, and ammonium acetate.
- ❑ Using an automatic delivery pipette add the water.
- ❑ Clamp the vessel over a stirring hotplate and stir the contents to dissolve the solids.
- ❑ Seal the reaction vessel with a cap according to the microwave manufacturer's recommendations.
- ❑ Place the sealed reaction vessel in the microwave cavity.
- ❑ Program the microwave unit to heat the vessel contents to 100°C over a 3-minute ramp period and then hold at this temperature for 15 minutes.
- ❑ After the heating step is completed, allow the contents of the reaction vessel to cool to 50°C or below before removing it from the microwave cavity.

## Isolate the Product

- ❑ Prepare two ice baths by taking crushed ice and placing it in two beakers.
- ❑ Carefully open the reaction vessel.
- ❑ Remove the stirring bar with a magnetic retrieving wand or a pair of tweezers.

- ❑ Cool the contents of the reaction vessel by placing the tube in one of the ice baths for 5 minutes.
- ❑ While the solution is cooling set up a vacuum filtration system with a Hirsch funnel, side-arm flask, rubber collar, and a length of rubber vacuum tubing.
- ❑ In the other ice bath, cool some water (2–4 mL) in a 25-mL Erlenmeyer flask for washing the precipitate after filtration.
- ❑ Connect the filtration system to a vacuum and place the correct size filter paper in the funnel.
- ❑ Wet the filter paper with a few drops of the cold water and start the vacuum to seal the filter paper in place.
- ❑ Filter the reaction mixture by pouring the contents into the funnel; transfer as much solid as possible.
- ❑ Rinse the reaction vessel with cold water (1 mL) and add the washings to the filter funnel.
- ❑ Rinse the precipitate on the filter with additional cold water (1 mL).
- ❑ Allow the precipitate to dry on the funnel for 10 minutes.
- ❑ Transfer the solid to a larger piece of filter paper to dry completely.
- ❑ Weigh the dry product and calculate the yield and percentage yield.

## Characterize the Product

- ❑ Obtain a $^{195}$Pt-NMR spectrum in $D_2O$ if instructed to do so.

## PROCEDURE FOR USE IN A MULTIMODE MICROWAVE UNIT

### PREPARATION OF *CIS*-DIAMMINEDICHLOROPLATINUM(II) (CISPLATIN)

$$K_2PtCl_4 + KCl + (NH_4)^+(CH_3CO_2)^- \xrightarrow{MW} cis\text{-}PtCl_2(NH_3)_2$$

**Caution:** Platinum salts are toxic and should be handled with care. The use of goggles with side-shields, lab coats, and gloves is considered minimum and non-discretionary safety practice in the laboratory.

---

### Table of Reagents and Physical Constants

| Reagent | Equiv. | FW | mmol | Mass (mg) | Vol. (mL) | mp/bp°C |
|---|---|---|---|---|---|---|
| Potassium tetrachloroplatinate(II) $K_2PtCl_4$ | 1 | 415 | 0.12 | 50 | — | 265 |
| Potassium chloride KCl | 5.6 | 74.5 | 0.52 | 50 | — | 770 |
| Ammonium acetate $C_2H_7NO_2$ | 4.3 | 77.0 | 0.51 | 40 | — | 114 |
| Water $H_2O$ | — | — | — | — | 2.5 | 100 |

---

### PERFORM THE REACTION

- ❑ In a 25-mL glass microwave reaction vessel, place a magnetic stir bar, the potassium tetrachloroplatinate, potassium chloride, and ammonium acetate.
- ❑ Using an automatic delivery pipette add the water.
- ❑ Clamp the vessel over a stirring hotplate and stir the contents to dissolve the solids.
- ❑ Seal the reaction vessel with a cap according to the microwave manufacturer's recommendations.
- ❑ Place the sealed reaction vessel in the carousel, noting the vessel's position number and ensuring that vessels are evenly spaced around the carousel.
- ❑ When all the group's reaction vessels are in place, load the carousel into the microwave cavity.
- ❑ If provided by the manufacturer, connect the temperature probe to the control vessel.
- ❑ Program the microwave unit to heat the vessel contents to 100°C over a 3-minute ramp period and then hold at this temperature for 15 minutes.
- ❑ After the heating step is completed, allow the contents of the reaction vessel to cool to 50°C or below before removing it from the microwave cavity.

### Isolate the Product

- ❑ Prepare two ice baths by taking crushed ice and placing it in two beakers.
- ❑ Carefully open the reaction vessel.
- ❑ Remove the stirring bar with a magnetic retrieving wand or a pair of tweezers.
- ❑ Cool the contents of the reaction vessel by placing the tube in one of the ice baths for 5 minutes.
- ❑ While the solution is cooling set up a vacuum filtration system with a Hirsch funnel, side-arm flask, rubber collar, and a length of rubber vacuum tubing.
- ❑ In the other ice bath, cool some water (4–8 mL) in a 25-mL Erlenmeyer flask for washing the precipitate after filtration.
- ❑ Connect the filtration system to a vacuum and place the correct size filter paper in the funnel.
- ❑ Wet the filter paper with a few drops of the cold water and start the vacuum to seal the filter paper in place.
- ❑ Filter the reaction mixture by pouring the contents into the funnel; transfer as much solid as possible.
- ❑ Rinse the reaction vessel with cold water (1–2 mL) and add the washings to the filter funnel.
- ❑ Rinse the precipitate on the filter with additional cold water (1–2 mL).
- ❑ Allow the precipitate to dry on the funnel for 10 minutes.
- ❑ Transfer the solid to a large piece of filter paper to dry completely.
- ❑ Weigh the dry product and calculate the yield and percentage yield.

### Characterize the Product

- ❑ Obtain a $^{195}$Pt-NMR spectrum in $D_2O$ if instructed to do so.

## QUESTIONS

1. Why is there a propensity for square planar geometry over tetrahedral in the case of metals with a d-electron configuration of $d^8$?
2. Do the following coordination complexes exhibit isomerism? If so, draw the different isomers and explain what kind of isomerism they exhibit.

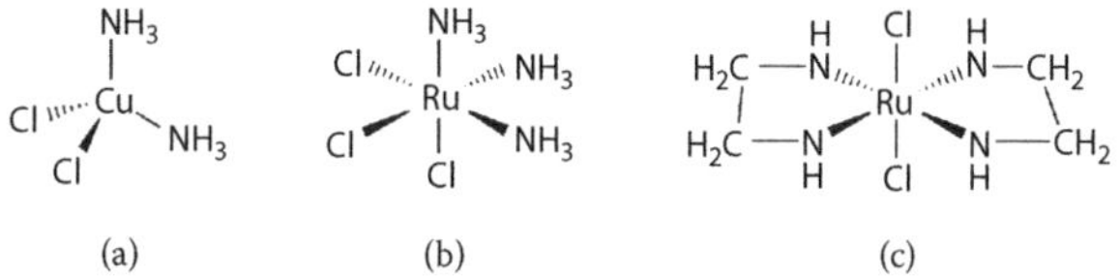

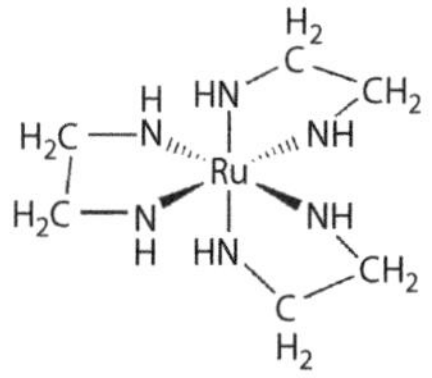

(a) (b) (c) (d)

3. By researching the literature, explain how cisplatin binds to DNA in cancer cells and, in so doing, stops cell reproduction.

# 18 Preparation of a Palladium Complex

## *Bis(triphenylphosphine) Palladium(II) Dichloride**

**LEARNING GOALS**

- To prepare a phosphorus-containing coordination complex
- To understand the concepts of isomerism in transition metal complexes
- To use multinuclear NMR spectroscopy

## INTRODUCTION

Metals bind to a wide range of atoms and molecules, these being generically termed ligands. Metal complexes bearing four ligands can adopt one of two possible geometries, namely square planar or tetrahedral. Square planar coordination is rare except in the case of metals with a d-electron configuration of $d^8$. This includes complexes of nickel, palladium, and platinum when the oxidation state of the metal is +2. When all the ligands around the central metal atom are the same, there is only one arrangement. However, when there are two different ligands attached to the metal and where there are two of each ligand, there are two possible isomeric structures. Looking at the complex MA2B2, where A and B are ligands attached to the central metal atom M, when both A ligands are next to each other the geometry is termed *cis*. When the two A ligands are opposite each other, the geometry is termed *trans*.

A, B, M, A, B

*cis isomer*

B, A, M, A, B

*trans isomer*

Many palladium complexes serve as active catalysts for important reactions in organic chemistry. This was reflected in the 2010 Nobel Prize in Chemistry which

* Experimental procedure developed by Benjamin Giorgi and Trevor Trueman, University of Connecticut.

was awarded to three chemists for their pioneering work in this field, specifically the use of palladium complexes in cross-coupling reactions. A cross-coupling reaction forms a carbon–carbon bond between two hydrocarbon fragments. Usually an organic compound containing a main-group atom RM (R = organic fragment, M = main group center) reacts with an alkyl or aryl halide R′X (R′ = organic fragment, X = halogen atom) in the presence of a palladium catalyst, forming a new carbon–carbon bond in the product R–R′. A range of different classes of main-group compounds can be used as coupling partners and the respective cross-coupling reactions are often named after the chemist that discovered them.

$$R{-}M \quad + \quad R'{-}X \xrightarrow{[Pd]} R{-}R' \qquad \textit{cross-coupling reaction}$$

*R = organic molecule*
*M = main-group metal fragment*
*R-B(OH)$_2$ Suzuki coupling*
*R-SnBu$_3$ Stille coupling*
*R-MgX Kumada coupling*
*R-ZnX Negishi coupling*

*R′ = organic molecule*
*X = halogen atom*

Of the wide range of palladium complexes used as catalysts for cross-coupling reactions, many contain phosphine ligands. Phosphine ligands have the general formula $PR_3$ where R = alkyl, aryl, H. They have a pyramidal structure, like ammonia. They are relatively easy to make and their steric and electronic properties can be tuned by varying the nature of the groups attached to the central phosphorus atom. Some phosphines are air-sensitive, being readily oxidized to the corresponding phosphine oxide $O = PR_3$. At one extreme is the parent compound phosphine $PH_3$, which is pyrophoric (catches fire in air). At the other end, triphenylphosphine, $PPh_3$, is relatively air stable and is one of the most widely used phosphines both as a ligand and also in organic transformations such as Wittig reactions.

*generic formula of a phosphine* *triphenylphosphine*

Palladium chloride reacts with triphenylphosphine to form a square planar complex bearing two chloride ligands and two triphenylphosphine ligands. Both *cis* and *trans* isomers are known. This complex, $PdCl_2(PPh_3)_2$, is widely used in cross-coupling reactions and is also an intermediate in the preparation of $Pd(PPh_3)_4$, another widely used palladium catalyst.

$$PdCl_2 + 2\,PPh_3 \longrightarrow PdCl_2(PPh_3)_2$$

$$PdCl_2(PPh_3)_2 + 2\,PPh_3 + 2.5\,N_2H_4 \longrightarrow Pd(PPh_3)_4 + 0.5\,N_2 + 2(N_2H_5)^+Cl^-$$

## PROCEDURE FOR USE IN A MONOMODE MICROWAVE UNIT

$$PdCl_2 + 2\,PPh_3 \xrightarrow{MW} PdCl_2(PPh_3)_2$$

**Caution:** Palladium salts are toxic and should be handled with care. The use of goggles with side-shields, lab coats, and gloves is considered minimum and non-discretionary safety practice in the laboratory.

### Table of Reagents and Physical Constants

| Reagent | Equiv. | FW | mmol | Mass (mg) | Density (g/mL) | Vol. (mL) | mp/bp°C |
|---|---|---|---|---|---|---|---|
| Palladium(II) chloride $PdCl_2$ | 1 | 177 | 0.14 | 25 | — | — | 500 (dec.) |
| Triphenylphosphine $C_{18}H_{15}P$ | 6 | 262 | 0.70 | 185 | — | — | 80 |
| Dimethyl sulfoxide $(CH_3)_2SO$ | — | — | — | — | 1.10 | 2.0 | 189 |
| Water $H_2O$ | — | — | — | — | 1.00 | 0.2 | 100 |

### Perform the Reaction

- ❑ In a 10-mL glass microwave reaction vessel, place a magnetic stir bar, the palladium(II) chloride, and triphenylphosphine.
- ❑ Using an automatic delivery pipette add the dimethyl sulfoxide and water.
- ❑ Clamp the vessel over a stirring hotplate and stir the contents to dissolve the solids.
- ❑ Seal the reaction vessel with a cap according to the microwave manufacturer's recommendations.
- ❑ Place the sealed reaction vessel in the microwave cavity.
- ❑ Program the microwave unit to heat the vessel contents to 140°C over a 3-minute ramp period and then hold at this temperature for 10 minutes.
- ❑ After the heating step is completed, allow the contents of the reaction vessel to cool to 50°C or below before removing it from the microwave cavity.

### Isolate the Product

- ❑ Prepare two ice baths by taking crushed ice and placing it in two beakers.
- ❑ Carefully open the reaction vessel.
- ❑ Remove the stirring bar with a magnetic retrieving wand or a pair of tweezers.
- ❑ Cool the contents of the reaction vessel by placing the tube in one of the ice baths for 5 minutes.

- ❑ While the solution is cooling set up a vacuum filtration system with a Hirsch funnel, side-arm flask, rubber collar, and a length of rubber vacuum tubing.
- ❑ In the other ice bath, cool some diethyl ether (2–4 mL) in a 25-mL Erlenmeyer flask for washing the precipitate after filtration.
- ❑ Connect the filtration system to a vacuum and place the correct size filter paper in the funnel.
- ❑ Wet the filter paper with a few drops of the cold diethyl ether and start the vacuum to seal the filter paper in place.
- ❑ Filter the reaction mixture by pouring the contents into the funnel; transfer as much solid as possible.
- ❑ Rinse the reaction vessel with cold diethyl ether (1 mL) and add the washings to the filter funnel.
- ❑ Rinse the precipitate on the filter with additional cold diethyl ether (1 mL).
- ❑ Allow the precipitate to dry on the funnel for 10 minutes.
- ❑ Transfer the solid to a larger piece of filter paper to dry completely.
- ❑ Weigh the dry product and calculate the yield and percentage yield.

### Characterize the Product

- ❑ Obtain an IR spectrum of the product.
- ❑ Obtain $^{1}H$- and $^{31}P$-NMR spectra in $CDCl_3$ if instructed to do so.

## PROCEDURE FOR USE IN A MULTIMODE MICROWAVE UNIT

$$PdCl_2 + 2\,PPh_3 \xrightarrow{MW} PdCl_2(PPh_3)_2$$

**Caution:** Palladium salts are toxic and should be handled with care. The use of goggles with side-shields, lab coats, and gloves is considered minimum and non-discretionary safety practice in the laboratory.

### Table of Reagents and Physical Constants

| Reagent | Equiv. | FW | mmol | Mass (mg) | Density (g/mL) | Vol. (mL) | mp/bp°C |
|---|---|---|---|---|---|---|---|
| Palladium(II) chloride $PdCl_2$ | 1 | 177 | 0.42 | 75 | — | — | 500 (dec.) |
| Triphenylphosphine $C_{18}H_{15}P$ | 6 | 262 | 2.10 | 555 | — | — | 80 |
| Dimethyl sulfoxide $(CH_3)_2SO$ | — | — | — | — | 1.10 | 6.0 | 189 |
| Water $H_2O$ | — | — | — | — | 1.00 | 0.6 | 100 |

### Perform the Reaction

- ❑ In a 25-mL glass microwave reaction vessel, place a magnetic stir bar, the palladium(II) chloride, and triphenylphosphine.
- ❑ Using an automatic delivery pipette add the dimethyl sulfoxide and water.
- ❑ Clamp the vessel over a stirring hotplate and stir the contents to dissolve the solids.
- ❑ Seal the reaction vessel with a cap according to the microwave manufacturer's recommendations.
- ❑ Place the sealed reaction vessel in the carousel, noting the vessel's position number and ensuring that vessels are evenly spaced around the carousel.
- ❑ When all the group's reaction vessels are in place, load the carousel into the microwave cavity.
- ❑ If provided by the manufacturer, connect the temperature probe to the control vessel.
- ❑ Program the microwave unit to heat the vessel contents to 140°C over a 10-minute ramp period and then hold at this temperature for 15 minutes.
- ❑ After the heating step is completed, allow the contents of the reaction vessel to cool to 50°C or below before removing it from the microwave cavity.

### Isolate the Product

- ❑ Prepare two ice baths by taking crushed ice and placing it in two beakers.
- ❑ Carefully open the reaction vessel.
- ❑ Remove the stirring bar with a magnetic retrieving wand or a pair of tweezers.
- ❑ Cool the contents of the reaction vessel by placing the tube in one of the ice baths for 5 minutes.
- ❑ While the solution is cooling set up a vacuum filtration system with a Hirsch funnel, side-arm flask, rubber collar, and a length of rubber vacuum tubing.
- ❑ In the other ice bath, cool some diethyl ether (8–10 mL) in a 25-mL Erlenmeyer flask for washing the precipitate after filtration.
- ❑ Connect the filtration system to a vacuum and place the correct size filter paper in the funnel.
- ❑ Wet the filter paper with a few drops of the cold diethyl ether and start the vacuum to seal the filter paper in place.
- ❑ Filter the reaction mixture by pouring the contents into the funnel; transfer as much solid as possible.
- ❑ Rinse the reaction vessel with cold diethyl ether (2 mL) and add the washings to the filter funnel.
- ❑ Rinse the precipitate on the filter with additional cold diethyl ether (2 mL).
- ❑ Allow the precipitate to dry on the funnel for 10 minutes.
- ❑ Transfer the solid to a larger piece of filter paper to dry completely.
- ❑ Weigh the dry product and calculate the yield and percentage yield.

### Characterize the Product

- ❑ Obtain an IR spectrum of the product.
- ❑ Obtain $^{1}H$- and $^{31}P$-NMR spectra in $CDCl_3$ if instructed to do so.

## QUESTIONS

1. Which isomer of the product do you think you have made (*cis* or *trans*)? Is there a method you could use to determine this experimentally?
2. What is the oxidation state of palladium in $PdCl_2(PPh_3)_2$ and $Pd(PPh_3)_4$? What does this tell you about the role of hydrazine ($N_2H_4$) in the preparation of $Pd(PPh_3)_4$ from $PdCl_2(PPh_3)_2$?
3. By researching the literature, find one example of the use of each of $PdCl_2(PPh_3)_2$ and $Pd(PPh_3)_4$ in a cross-coupling reaction.

# 19 Coordination of an Aromatic Ring to a Metal

## *Preparation of an Arene Chromium Tricarbonyl Complex**

**LEARNING GOALS**

- To prepare an organometallic complex
- To purify a metal complex by microscale flash column chromatography
- To understand the implications of coordinating an aromatic ring to a metal

## INTRODUCTION

Aromatic rings, also known as arenes are organic compounds that contain a conjugated planar ring system with delocalized π-electron clouds. The rings are electron rich and the delocalization brings enhanced stability, meaning that the chemistry of arenes is dominated by electrophilic substitution reactions such as nitration or Friedel–Crafts alkylation and not addition reactions as traditionally seen with alkenes.

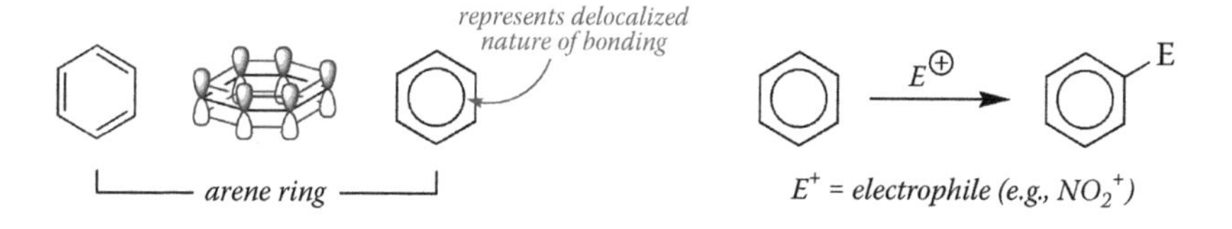


The reactivity of arenes can be dramatically altered by coordination to a metal compound. Metals can donate electron density into the π*-antibonding orbitals of the arene, this reducing the bond-order of the carbon–carbon bonds in the ring.

* Procedure developed by Trevor Trueman and Christopher Lee, University of Connecticut.

Arene chromium tricarbonyl complexes contain an arene ring bound to a central chromium atom and also bear three carbon monoxide groups (called carbonyls). The electron-withdrawing effect of the chromium tricarbonyl $Cr(CO)_3$ unit makes the arene ring prone to nucleophilic attack, stabilizes negative charges in benzylic positions, and activates carbon–halogen bonds for cross-coupling reactions such as Suzuki and Heck couplings. In addition, the $Cr(CO)_3$ group effectively blocks one face of the ring, this meaning that chemistry can be performed in a stereospecific manner (i.e., on the other face of the ring).

crabonyl groups

*arene chromium tricarbonyl complex*

increased acidity of ring hydrogens

stabilization of benzylic positive and negative charges

nucleophilic addition to ring

steric directing group for selective reactions on substituents

The chromium atom is bound to all six carbons of the ring in arene chromium tricarbonyl complexes. The term hapticity is used to describe how many atoms in a coordinated group are attached to the central atom. Hapticity of a ligand is indicated by the Greek character "eta", η. A superscripted number following the symbol denotes the number of coordinated atoms of the group that are bound to the metal. Thus, the arene group in arene chromium tricarbonyl complexes is termed hapto-6 or $\eta^6$.

Arene chromium carbonyl complexes are traditionally prepared by heating chromium hexacarbonyl $Cr(CO)_6$ either in the arene as the solvent or in a high boiling solvent containing the arene. The products are somewhat air sensitive in solution but are stable to air in the solid state and can be stored for long periods provided that they are kept out of the light.

## PROCEDURE FOR USE IN A MONOMODE MICROWAVE UNIT

### PREPARATION OF TOLUENE CHROMIUM TRICARBONYL $[(\eta^6\text{–}C_6H_5CH_3)Cr(CO)_3]$

$Cr(CO)_6$ + toluene $\xrightarrow{MW}$ $(\eta^6\text{–}C_6H_5CH_3)Cr(CO)_3$

**Caution:** Chromium hexacarbonyl is toxic and should be handled with care. The use of goggles with side-shields, lab coats, and gloves is considered minimum and nondiscretionary safety practice in the laboratory.

#### Table of Reagents and Physical Constants

| Reagent | Equiv. | FW | mmol | Mass (mg) | Density (g/mL) | Vol. (mL) | mp/bp°C |
|---|---|---|---|---|---|---|---|
| Chromium hexacarbonyl $C_6O_6Cr$ | 1 | 220 | 0.45 | 100 | — | — | 150 |
| Toluene $C_7H_8$ | 21 | 92 | 9.41 | 867 | 0.867 | 1.0 | 110 |
| Tetrahydrofuran $C_4H_8O$ | — | — | — | — | — | 1.0 | 66 |

### PERFORM THE REACTION

- ❑ In a 10-mL glass microwave reaction vessel, place a magnetic stir bar and the chromium hexacarbonyl.
- ❑ Using an automatic delivery pipette add the toluene and tetrahydrofuran.
- ❑ Seal the reaction vessel with a cap according to the microwave manufacturer's recommendations.
- ❑ Place the sealed reaction vessel in the microwave cavity.
- ❑ Program the microwave unit to heat the vessel contents to 160°C, using an initial microwave power of 250 W, and hold at this temperature for 60 minutes.
- ❑ After the heating step is completed, allow the contents of the reaction vessel to cool to 50°C or below before removing it from the microwave cavity.

### ISOLATE THE PRODUCT

- ❑ While the reaction mixture is cooling, plug a 5.75-mm disposable glass pipette with a small piece of glass wool.
- ❑ Add to the pipette, silica gel to a height of 7 cm.
- ❑ Add a layer of sand (5 mm) to the top of the silica in the pipette.

- ❑ Carefully open the reaction vessel.
- ❑ Pour the contents of the reaction vessel into a 25-mL round-bottom flask.
- ❑ Rinse the reaction vessel with 1-mL tetrahydrofuran, placing the washings into the round-bottom flask containing the product mixture.
- ❑ Place the round-bottom flask on a rotary evaporator and remove the liquid under reduced pressure.
- ❑ Once almost all the liquid has been removed, take the flask off the rotary evaporator and re-dissolve any solid using a minimal volume of tetrahydrofuran.
- ❑ Add the concentrated product solution to the top of the column.
- ❑ Once all the product has entered the column, add 10% ethyl acetate in pentane solution to the top of the column.
- ❑ Collect the eluting liquid in a tared clean round-bottom flask.
- ❑ Stop collecting once all the green-colored material has passed through the column and into the round bottom flask.
- ❑ Remove the solvent under reduced pressure until a constant weight is observed.
- ❑ Re-weigh the flask containing the yellow/green solid.
- ❑ Calculate the yield and percent yield.

## Characterize the Product

- ❑ Obtain an infrared spectrum..
- ❑ Obtain a $^{1}$H-NMR spectrum in $CDCl_3$.

## PROCEDURE FOR USE IN A MULTIMODE MICROWAVE UNIT

### PREPARATION OF TOLUENE CHROMIUM TRICARBONYL [($\eta^6$–$C_6H_5CH_3$)$Cr(CO)_3$]

$Cr(CO)_6$ + toluene $\xrightarrow{MW}$ ($\eta^6$-toluene)$Cr(CO)_3$

**Caution:** Chromium hexacarbonyl is toxic and should be handled with care. The use of goggles with side-shields, lab coats, and gloves is considered minimum and nondiscretionary safety practice in the laboratory.

---

### Table of Reagents and Physical Constants

| Reagent | Equiv. | FW | mmol | Mass (mg) | Density (g/mL) | Vol. (mL) | mp/bp°C |
|---|---|---|---|---|---|---|---|
| Chromium hexacarbonyl $C_6O_6Cr$ | 1 | 220 | 0.90 | 200 | — | — | 150 |
| Toluene $C_7H_8$ | 31.5 | 92 | 28.23 | 2601 | 0.867 | 3.0 | 110 |
| Tetrahydrofuran $C_4H_8O$ | — | — | — | — | — | 2.0 | 66 |

---

### PERFORM THE REACTION

- ❑ In a 25-mL glass microwave reaction vessel, place a magnetic stir bar and the chromium hexacarbonyl.
- ❑ Using an automatic delivery pipette add the toluene and tetrahydrofuran.
- ❑ Seal the reaction vessel with a cap according to the microwave manufacturer's recommendations.
- ❑ Place the sealed reaction vessel in the carousel, noting the vessel's position number and ensuring that vessels are evenly spaced around the carousel.
- ❑ When all the group's reaction vessels are in place, load the carousel into the microwave cavity.
- ❑ If provided by the manufacturer, connect the temperature probe to the control vessel.
- ❑ Program the microwave unit to heat the vessel contents to 160°C, using an initial microwave power of 900 W, and hold at this temperature for 60 minutes.
- ❑ After the heating step is completed, allow the contents of the reaction vessel to cool to 50°C or below before removing it from the microwave cavity.

### Isolate the Product

- ❑ While the reaction mixture is cooling, plug a 5.75-mm disposable glass pipette with a small piece of glass wool.
- ❑ Add to the pipette, silica gel to a height of 7 cm.
- ❑ Add a layer of sand (5 mm) to the top of the silica in the pipette.
- ❑ Carefully open the reaction vessel.
- ❑ Pour the contents of the reaction vessel into a 25-mL round-bottom flask,
- ❑ Rinse the reaction vessel with 1 mL tetrahydrofuran, placing the washings into the round-bottom flask containing the product mixture.
- ❑ Place the round-bottom flask on a rotary evaporator and remove the liquid under reduced pressure.
- ❑ Once almost all the liquid has been removed, take the flask off the rotary evaporator and re-dissolve any solid using a minimal volume of tetrahydrofuran.
- ❑ Add the concentrated product solution to the top of the column.
- ❑ Once all the product has entered the column, add 10% ethyl acetate in a pentane solution to the top of the column.
- ❑ Collect the eluting liquid in a tared clean round-bottom flask.
- ❑ Stop collecting once all the green-colored material has passed through the column and into the round bottom flask.
- ❑ Remove the solvent under reduced pressure until a constant weight is observed.
- ❑ Re-weigh the flask containing the yellow/green solid.
- ❑ Calculate the yield and percent yield.

### Characterize the Product

- ❑ Obtain an infrared spectrum.
- ❑ Obtain a $^{1}$H-NMR spectrum in $CDCl_3$.

## QUESTIONS

1. How do the signals in the $^1$H-NMR for the toluene ring in the chromium complex compare with those for free toluene? What does this tell you about the effects of coordination on the bonding in the ring?
2. What is the point group of arene chromium tricarbonyl complexes?
3. By researching the literature, find one practical application of arene chromium tricarbonyl complexes as reagents in organic synthesis.
4. Chromium is situated above molybdenum (Mo) and tungsten (W) in the periodic table. Although it is possible to prepare arene complexes with Mo and W from the hexacarbonyl starting materials [$Mo(CO)_6$ and $W(CO)_6$ respectively], it is much harder to do than with chromium. Why is this?

# 20 Determination of an Empirical Formula

## *Zinc Bromide**

**LEARNING GOALS**

- To determine an empirical formula experimentally
- To use the concept of a limiting reagent

## INTRODUCTION

The empirical formula of a compound is the simplest whole number ratio of the atoms that are present in the compound. It is often determined from the elemental analysis of the compound. There is a variety of quantitative and qualitative methods to determine the elements present in the original sample. For example, the most common for organic chemicals is combustion analysis. This technique burns the organic material (which is comprised mostly of carbon, hydrogen, nitrogen, oxygen, and halogens) in excess oxygen. The mass of carbon dioxide, water, and nitric oxide that is produced is used to determine the percent composition of carbon, hydrogen, and nitrogen, respectively, in the original compound. This method is destructive to the original sample.

An alternative technique for determining the empirical formula of diatomic compounds is to synthesize the compound from a known quantity of one of the elements, the limiting reagent, and a large excess of the other element. The limiting reagent is the starting material that is completely consumed during the reaction. After the reaction the amount of unreacted element that is in excess is determined. The difference between the initial mass of excess reagent before the reaction and the mass of the reagent unreacted after the reaction is equal to the mass of reagent consumed during the reaction. The stoichiometry of the reaction, which relates the moles of reactants to the moles of products in a balanced chemical equation, can be used to determine the empirical formula.

$$A + nB \longrightarrow AB_y + zB \qquad \textit{mass of B in } AB_y = \textit{mass of nB} - \textit{mass of zB}$$

↖ *limiting reagent* (A)

* Experimental procedure developed by Dr. Javier Horta and David Forman, Merrimack College.

In this experiment, a known mass of the limiting reagent, bromine, is combined with an excess of granular zinc to produce the inorganic salt, zinc bromide. Bromine is a diatomic element ($Br_2$) that is a very toxic red liquid which vaporizes readily at room temperature. A safer equivalent to bromine is the solid pyridinium tribromide. Each equivalent of this compound delivers one equivalent of reactive $Br_2$. Therefore, each mole of pyridinium tribromide that is consumed during the reaction results in two moles of bromine atoms (or one mole of bromine molecules) in the product.

A known mass of zinc is the reagent in excess. Upon completion of the reaction the unreacted zinc is isolated. The mass of zinc consumed during the reaction is equal to the difference between the initial mass of zinc and the mass of zinc isolated at the end. To calculate the empirical formula the mass of each element consumed must be converted to the mole equivalent by dividing the mass by the molar mass of the element (g/mol). The empirical formula can then be calculated.



*pyridinium tribromide*

$$mZn + nBr_2 \longrightarrow Zn_pBr_{2n} + yZn$$

*m and n are known (they are the amounts of the two starting materials added)*
*y is determined by analysis at the end of the reaction*
$p = y - m$

## PROCEDURE FOR USE IN A MONOMODE MICROWAVE UNIT

**Caution:** Pyridinium tribromide is a safer reagent than elemental bromine but is still corrosive and a lachrymator. It should be weighed and used in a fume hood. The use of goggles with side-shields, lab coats, and gloves is considered minimum and nondiscretionary safety practice in the laboratory.

### Table of Reagents and Physical Constants

| Reagent | Equiv. | FW | mmol | Mass (mg) | Vol. (mL) | mp/bp°C |
|---|---|---|---|---|---|---|
| Zinc<br>Zn | 4.9 | 65.4 | 4.6 | 300 | — | — |
| Pyridinium tribromide<br>$C_5H_6Br_3N$ | 1 | 319.8 | 0.94 | 300 | — | 127–133 |
| Water<br>$H_2O$ | — | — | — | — | 5.0 | 100 |

### Perform the Reaction (in Triplicate)

- ❑ In three 10-mL glass microwave reaction vessels place a magnetic stir bar.
- ❑ Label the reaction vessels so that they can be identified.
- ❑ Weigh and record the weight of each reaction vessel plus stirring bar to the nearest milligram (weight 1).
- ❑ Add the granular zinc to each vessel (300 mg per vessel).
- ❑ Weigh and record the weight of each reaction vessel, stirring bar, and zinc to the nearest milligram (weight 2).
- ❑ Add pyridinium hydrobromide perbromide to each reaction vessel (300 mg per vessel).
- ❑ Weigh and record the weight of each reaction vessel containing the stirring bar, zinc, and pyridinium hydrobromide perbromide to the nearest milligram (weight 3).
- ❑ Add the water to each vessel using a graduated cylinder (5 mL per vessel).
- ❑ Seal the first reaction vessel with a cap according to the microwave manufacturer's recommendations.
- ❑ Place the sealed reaction vessel in the microwave cavity.
- ❑ Program the microwave unit to heat the vessel contents to 130°C using an initial microwave power of 150 W and hold at this temperature for 15 minutes.
- ❑ After the heating step is completed, allow the contents of the reaction vessel to cool to 50°C or below before removing it from the microwave cavity.
- ❑ Repeat the heating step with each of the other two reaction vessels.

## Postreaction Procedure

- ❑ Carefully open the reaction vessel.
- ❑ Using a Pasteur pipette, add acetone (2 mL) to the vessel to help settle any solids.
- ❑ Gently shake the vessel to mix the solvent allowing the unreacted zinc to settle to the bottom.
- ❑ Decant the solution into a 50-mL Erlenmeyer flask leaving all the solids behind.
- ❑ Rinse the solids in the reaction vessel with water (3 mL).
- ❑ Add a small amount of acetone to settle the solids, if needed.
- ❑ Decant the solution into a 50-mL Erlenmeyer flask leaving all the solids behind.
- ❑ Repeat the rinse two times.
- ❑ Rinse the solids with acetone (3 mL).
- ❑ Decant the acetone, leaving the solids in the reaction vessel.
- ❑ Repeat the acetone rinse two times.
- ❑ Dry the reaction vessel with the solids in an oven or with a heat gun.
- ❑ Weigh and record the weight of the vessel with the unreacted zinc solids (weight 4).
- ❑ Repeat the postreaction procedure with each of the other two reaction vessels.

## Determination of the Empirical Formula

For each vessel:

- ❑ Calculate the initial amount of zinc before the reaction (weight 2 – weight 1).
- ❑ Calculate the amount of unreacted zinc (weight 4 – weight 1).
- ❑ Calculate the amount of zinc consumed (initial amount – unreacted zinc).
- ❑ Determine the moles of zinc that reacted (zinc consumed/molar mass of zinc).
- ❑ Calculate the amount of pyridinium tribromide added (weight 3 – weight 2).
- ❑ Determine the number of moles of pyridinium tribromide added to each vessel (mass of pyridinium tribromide/molar mass of pyridinium tribromide).
- ❑ Determine the empirical formula for zinc bromide (ratio of the number of moles of zinc and pyridinium tribromide).

## PROCEDURE FOR USE IN A MULTIMODE MICROWAVE UNIT

**Caution:** Pyridinium tribromide is a safer reagent than elemental bromine but is still corrosive and a lachrymator. It should be weighed and used in a fume hood. The use of goggles with side-shields, lab coats, and gloves is considered minimum and nondiscretionary safety practice in the laboratory.

### Table of Reagents and Physical Constants

| Reagent | Equiv. | FW | mmol | Mass (mg) | Vol. (mL) | mp/bp°C |
|---|---|---|---|---|---|---|
| Zinc<br>Zn | 4.9 | 65.4 | 4.60 | 300 | — | — |
| Pyridinium tribromide<br>$C_5H_6Br_3N$ | 1 | 319.8 | 0.94 | 300 | — | 127–133 |
| Water<br>$H_2O$ | — | — | — | — | 5.0 | 100 |

### Perform the Reaction (in Triplicate)

- ❑ In three 25-mL glass microwave reaction vessels place a magnetic stir bar.
- ❑ Label the reaction vessels so that they can be identified.
- ❑ Weigh and record the weight of each reaction vessel plus stirring bar to the nearest milligram (weight 1).
- ❑ Add the granular zinc to each vessel (300 mg per vessel).
- ❑ Weigh and record the weight of each reaction vessel, stirring bar, and zinc to the nearest milligram (weight 2).
- ❑ Add pyridinium hydrobromide perbromide to each reaction vessel (300 mg per vessel).
- ❑ Weigh and record the weight of each reaction vessel containing the stirring bar, zinc, and pyridinium hydrobromide perbromide to the nearest milligram (weight 3).
- ❑ Add the water to each vessel using a graduated cylinder (5 mL per vessel).
- ❑ Seal the first reaction vessel with a cap according to the microwave manufacturer's recommendations.
- ❑ Place the sealed reaction vessels in the carousel, noting the vessel's position numbers and ensuring that vessels are evenly spaced around the carousel.
- ❑ When all the group's reaction vessels are in place, load the carousel into the microwave cavity.
- ❑ If provided by the manufacturer, connect the temperature probe to the control vessel.
- ❑ Program the microwave unit to heat the vessel contents to 130°C using an initial microwave power of 600 W and hold at this temperature for 15 minutes.

- ❑ After the heating step is completed, allow the contents of the reaction vessels to cool to 50°C or below before removing them from the microwave cavity.

## Postreaction Procedure

- ❑ Carefully open the reaction vessel.
- ❑ Using a Pasteur pipette, add acetone (2 mL) to the vessel to help settle any solids.
- ❑ Gently shake the vessel to mix the solvent allowing the unreacted zinc to settle to the bottom.
- ❑ Decant the solution into a 50-mL Erlenmeyer flask leaving all the solids behind.
- ❑ Rinse the solids in the reaction vessel with water (3 mL).
- ❑ Add a small amount of acetone to settle the solids, if needed.
- ❑ Decant the solution into a 50-mL Erlenmeyer flask leaving all the solids behind.
- ❑ Repeat the rinse two times.
- ❑ Rinse the solids with acetone (3 mL).
- ❑ Decant the acetone, leaving the solids in the reaction vessel.
- ❑ Repeat the acetone rinse two times.
- ❑ Dry the reaction vessel with the solids in an oven or with a heat gun.
- ❑ Weigh and record the weight of the vessel with the unreacted zinc solids (weight 4).
- ❑ Repeat the postreaction procedure with each of the other two reaction vessels.

## Determination of the Empirical Formula

For each vessel:

- ❑ Calculate the initial amount of zinc before the reaction (weight 2 – weight 1).
- ❑ Calculate the amount of unreacted zinc (weight 4 – weight 1).
- ❑ Calculate the amount of zinc consumed (initial amount – unreacted zinc).
- ❑ Determine the moles of zinc that reacted (zinc consumed/molar mass of zinc).
- ❑ Calculate the amount of pyridinium tribromide added (weight 3 – weight 2).
- ❑ Determine the number of moles of pyridinium tribromide added to each vessel (mass of pyridinium tribromide/molar mass of pyridinium tribromide).
- ❑ Determine the empirical formula for zinc bromide (ratio of the number of moles of zinc and pyridinium tribromide).

## QUESTIONS

1. What is the electron configuration of the elements zinc and bromine?
2. What common ionic charge would be expected for zinc and bromine?
3. What is the formula predicted for the ionic compound zinc bromide and how does this agree or disagree with the experimental data?
4. Determine the empirical formula of the organic compound with the following composition by mass: 48.0% C, 4.0% H, and 48.0% O.
5. What is the empirical formula for a compound containing 26.57% potassium, 35.36% chromium, and 38.07% oxygen?

# 21 Microwave-Assisted Extraction

## *Identification of the Major Flavor Components of Citrus Oil**

**LEARNING GOALS**

- To use microwave heating to extract compounds from a complex mixture
- To analyze mixtures of compounds using gas chromatography–mass spectrometry
- To understand the concept of headspace analysis

## INTRODUCTION

The essential oils of various fruits and spices are largely responsible for their aroma and flavor. Among the important volatile components found in the essential oils of citrus fruits are the terpenes, a group of unsaturated hydrocarbons composed of isoprenoid ($C_5H_8$) units. The monoterpenes each contain two isoprene units and are biosynthesized via the universal isoprenoid intermediate geranyl pyrophosphate (GPP). Among the major monoterpenes found in citrus fruits are α-pinene, β-pinene, limonene, myrcene, and linalool (a terpene alcohol), which can be extracted from the rind of the fruit. Differences in the occurrence and relative abundance of the monoterpenes in different fruits give rise to differences in taste and smell.

* Experimental procedure developed by Dr. K. C. Swallow, Victoria Hansen, and Lauren Viarengo, Merrimack College.

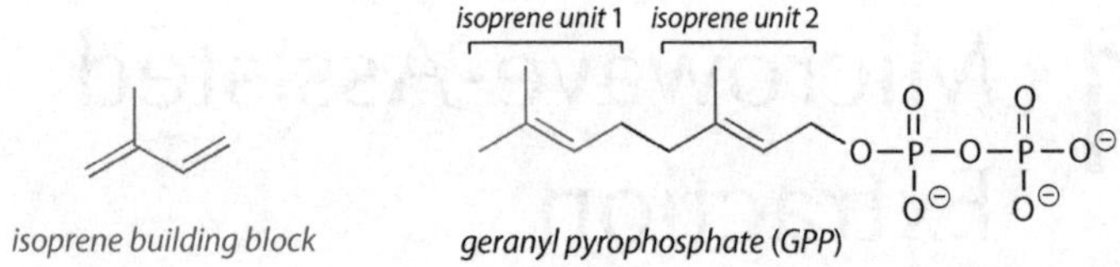

isoprene unit 1 isoprene unit 2

isoprene building block

geranyl pyrophosphate (GPP)

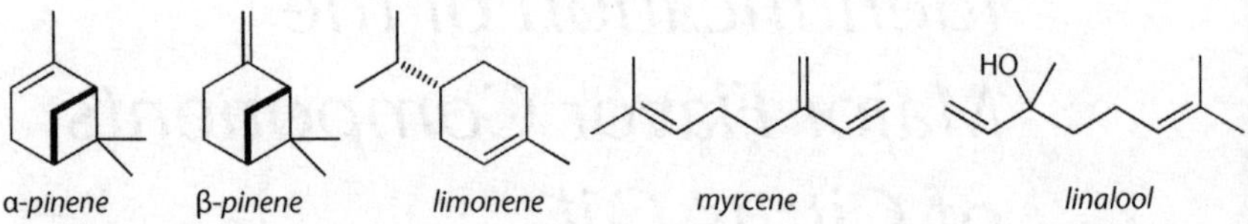

α-pinene β-pinene limonene myrcene linalool

Essential oils have been extracted from citrus peels using cold pressing, steam distillation, ultrasound-assisted extraction, and supercritical fluid extraction among other techniques. Microwave extraction into water is an attractive alternative that uses a simple apparatus with no organic solvent required and can be accomplished in under 20 minutes. Gas chromatography–mass spectrometry (GC/MS) can then be used to separate and identify the major components of the essential oils in the extract. Comparison of the retention times and mass spectra of the sample components and a set of known terpene standards allows for identification of the specific set of terpenes and determination of their relative abundance in citrus fruit extracts.

As water cannot be directly injected into a GC/MS apparatus, the terpenes must be extracted from the aqueous solution prior to analysis. Solid-phase microextraction (SPME) is a technique in which the organic compounds in a sample are adsorbed onto a coated fiber. The adsorption of the organics onto the fiber can be carried out in the aqueous phase, with the fiber inserted directly into an aqueous sample, or in the gas phase, with the fiber inserted into the headspace above the aqueous sample in a closed vial. In either case the fiber is then inserted into the injection port of the GC/MS where the volatile organic compounds are thermally desorbed and swept onto the column in the gas phase. Headspace analysis is especially well suited to the analysis of the volatile terpenes in citrus peel extracts.

## PROCEDURE FOR USE IN A MULTIMODE MICROWAVE UNIT

### MICROWAVE-ASSISTED EXTRACTION OF ESSENTIAL OILS FROM A SAMPLE OF CITRUS PEEL AND THEIR CHARACTERIZATION BY GAS CHROMATOGRAPHY–MASS SPECTROMETRY

**Caution:** Preparation of the sample involves the use of a sharp knife. Although the extracts are all natural products, the compounds can cause skin and eye irritation. The use of goggles with side-shields, lab coats, and gloves is considered minimum and nondiscretionary safety practice in the laboratory.

### Table of Reagents and Physical Constants

| Terpene Reference Compounds | FW | Density (g/mL) | bp (°C) |
|---|---|---|---|
| Limonene 1-methyl-4-prop-1-en-2-ylcyclohexene $C_{10}H_{16}$ | 136.24 | 0.840 | 175–176 |
| Linalool 3,7-dimethylocta-1,6-dien-3-ol $C_{10}H_{18}O$ | 154.25 | 0.858–0.867 | 194–197 |
| Myrcene 7-methyl-3-methylideneocta-1,6-diene $C_{10}H_{16}$ | 136.23 | 0.789–0.793 | 166–167 |
| (+)-α-Pinene (1R,5R)-4,7,7-trimethylbicyclo[3.1.1]hept-3-ene $C_{10}H_{16}$ | 136.23 | 0.858 | 155–156 |
| β-Pinene (1S,5S)-7,7-dimethyl-4-methylidenebicyclo[3.1.1]heptane $C_{10}H_{16}$ | 136.23 | 0.864–0.872 | 165–167 |
| γ-Terpinene 1-methyl-4-propan-2-ylcyclohexa-1,4-diene $C_{10}H_{16}$ | 136.23 | 0.841–0.845 | 181–183 |
| Terpinolene 1-methyl-4-propan-2-ylidenecyclohexene $C_{10}H_{16}$ | 136.23 | 0.872–0.882 | 183–185 |

### PREPARE THE SAMPLES

- ❑ Remove a ~3-$cm^2$ piece of peel from a citrus fruit (lemon, lime, orange, or grapefruit).
- ❑ Remove the pith by scraping with a spatula or knife.
- ❑ Using a sharp knife, cut the peel into strips approximately 2 mm × 5 mm in size.

### PERFORM THE EXTRACTION

- ❑ In a 25-mL glass microwave reaction vessel place a magnetic stir bar.
- ❑ Weigh ~0.5 g of the peel strips and place them into the microwave vessel.

- ❑ Add deionized water (5 mL) to the microwave vessel (the peel does not have to be completely submerged).
- ❑ Seal the reaction vessel with a cap according to the microwave manufacturer's recommendations.
- ❑ Place the sealed reaction vessel in the carousel, noting the vessel's position number and ensuring that vessels are evenly spaced around the carousel.
- ❑ When all the group's reaction vessels are in place, load the carousel into the microwave cavity.
- ❑ If provided by the manufacturer, connect the temperature probe to the control vessel.
- ❑ Program the microwave unit to heat the contents of the reaction vessels to 50°C over a 1-minute ramp period and then hold at this temperature for 17 minutes.
- ❑ After the heating step is completed, allow the contents of the reaction vessel to cool to 50°C or below before removing it from the microwave cavity.

### Prepare the Samples for Analysis

- ❑ Decant the liquid from the reaction vessel into a labeled 16-mL glass vial with a PTFE/SIL septum cap.
- ❑ Use a large-gauge syringe needle to make a hole in the septum and clamp the vial onto a ring stand.
- ❑ Load an SPME holder with a carboxen/polydimethylsiloxane (CAR/PDMS) fiber.
- ❑ With the fiber retracted, insert the needle through the hole in the septum. Leave a sufficient gap between the end of the needle and the liquid level for the fiber to be extruded only into the headspace above the level of the liquid and not into the liquid itself.
- ❑ Clamp the SPME holder in place above the vial. Extrude the fiber and leave it in the headspace for 5 minutes. Retract the fiber and remove the SPME holder from the vial.

### Set Up the GC/MS and Analyze the Terpene Reference Samples

- ❑ Equip the GC/MS with a 30 m × 0.25 mm × 0.25 μm 5% phenyl–95% dimethylpolysiloxane glass capillary column, helium carrier gas, a SPME inlet liner, and a low-bleed septum with an injection hole or a septumless inlet system.
- ❑ Set the inlet temperature to 300°C and use a 10:1 split ratio.
- ❑ Enter the temperature program (total run-time is 28 minutes):
    - ❑ Initial temperature 50°C.
    - ❑ Ramp at a rate of 5°C/minute to 175°C.
    - ❑ Hold 175°C for 3 minutes.

- ❑ Prepare approximately 100 μg/mL standards of the individual terpenes (α-pinene, β-pinene, myrcene, limonene, γ-terpinine, terpinolene, and linalool) by injecting 1 μL of the pure liquid into 10 mL of dichloromethane.
- ❑ Inject 1 μL of each solution into the GC/MS, allowing the program to reach completion before injecting the next.
- ❑ Record the retention time and generate the mass spectrum of each standard.

## Perform the GC/MS Analysis of Terpenes in the Headspace Sample

- ❑ Retract the fiber and remove the SPME holder from the sample vial.
- ❑ Insert the needle of the SPME holder into the injection port of the gas chromatograph.
- ❑ Extrude the SPME fiber and leave it in the injection port for about 2 minutes.
- ❑ Start the temperature program.
- ❑ Retract the fiber and remove the SPME holder from the injection port. Once it has returned to room temperature it can be used to collect the volatiles from another sample.
- ❑ Record the retention time and generate the mass spectrum for each peak in the chromatogram.
- ❑ Compare the data from the sample to the standards to identify each of the major components of the sample.

## QUESTIONS

1. What is the empirical formula for naturally occurring hydrocarbons called terpenes?
2. Except for linalool, all of the terpenes used in the experiment have the same molecular formula and formula weight. Look up the structures of these compounds. Are they stereoisomers or constitutional isomers?
3. How could you predict the relative gas chromatographic retention times for these compounds?
4. 1-[1,5-Dimethyl-4-hexen-1-yl]-1,2,3,4,4a,5,6,8a-octahydro-7-methyl-4-methylenenapthalene, shown below is a diterpene isolated from termite soldiers. Its structure is composed of four isoprene skeleton units. Identify the carbons that compose the four isoprene units within the overall structure.

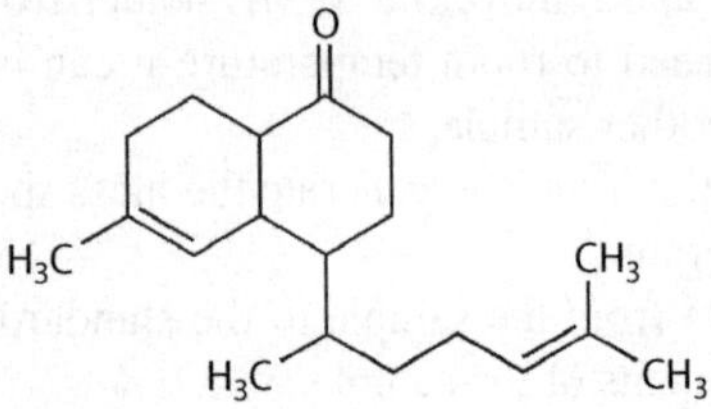

O
$H_3C$
$CH_3$
$H_3C$
$CH_3$

# 22 Microwave-Assisted Digestion of Dietary Supplements

## *Metal Analysis by Atomic Absorption Spectroscopy*[*]

**LEARNING GOALS**

- To use microwave heating to digest a sample
- To perform a metal analysis using atomic absorption spectroscopy
- To understand the difference between qualitative and quantitative analysis

## INTRODUCTION

Many people believe dietary supplements derived from natural products provide health benefits without the risks associated with synthetic products. Even naturally derived products, however, may contain unwanted ingredients such as metals arising either from the source itself or the manufacturing process. Some trace metals are potentially toxic at low concentrations. Unlike pharmaceuticals, dietary supplements are not regulated by the Food and Drug Administration (FDA) unless problems arise after they have been used by the public. The manufacturer is solely responsible for ensuring the safety of a product before it is marketed.

An extract derived from acai berries, the fruit of the acai palm found in Central and South America, is marketed as a powerful antioxidant with purported health benefits ranging from detoxification to promoting weight loss. It also is claimed to lower cholesterol and fight diseases such as arthritis and cancer. It is available both in food products and in tablets or capsules.

To analyze tablets or capsules for metals using atomic absorption (AA) or inductively coupled plasma (ICP) emission, it is necessary to dissolve the sample in solution. Wet ashing, the digestion of a sample using mineral acids to destroy

---

[*] Experimental procedure developed by Dr. K. C. Swallow and Victoria Hansen, Merrimack College.

Acai berries. (Image Copyright JBK, 2012. Used under license from Shutterstock.com)

the organic matrix and solubilize the metals as ions, requires several hours, produces noxious fumes, and is subject to losses from volatilization or entrainment. Microwave digestion in closed vessels is an attractive alternative to wet digestion. It provides better recovery of volatile elements, faster ashing due to higher pressures and temperatures, and less user exposure to fumes.

## PROCEDURE FOR USE IN A MULTIMODE MICROWAVE UNIT

**Caution:** Nitric acid and hydrochloric acid are strong acids that can cause serious burns if spilled on the skin. Transfer the acid to a glass graduated cylinder and pipette the required amount directly into the reaction tube. Flush any extra reagent down the sink with large amounts of water. If any sulfuric acid is accidentally spilled on the skin, immediately wash the area with large amounts of water to prevent a burn. The use of goggles with side-shields, lab coats, and gloves is considered minimum and nondiscretionary safety practice in the laboratory.

**Table of Reagents and Physical Constants**

| Reagent | Number | FW | Vol. (mL) |
|---|---|---|---|
| Acai supplement pills | 4 | — | — |
| Nitric acid (conc.) $HNO_3$ | — | 63.01 | 8.0 |
| Hydrochloric acid (conc.) HCl | — | 36.46 | 2.0 |

### Perform the Extraction

- ❑ In a 50-mL Teflon microwave reaction vessel place a magnetic stir bar.
- ❑ Grind three or four pills using a mortar and pestle or empty the contents of several capsules into a beaker to obtain a representative sample.
- ❑ Accurately weigh about 0.5 g of the powdered sample and transfer it to the Teflon reaction vessel.
- ❑ With extreme care, add the concentrated nitric and hydrochloric acids to the Teflon reaction vessel.
- ❑ Seal the reaction vessel with a cap according to the microwave manufacturer's recommendations.
- ❑ Place the sealed reaction vessel in the carousel, noting the vessel's position number and ensuring that vessels are evenly spaced around the carousel.
- ❑ When all the group's reaction vessels are in place, load the carousel into the microwave cavity.
- ❑ If provided by the manufacturer, connect the temperature probe to the control vessel.
- ❑ Program the microwave unit to heat the vessel contents using a three-step program.

| Stage | Power (W) | Stirring | Ramp Time (min) | Temperature (°C) | Hold Time (min) |
|---|---|---|---|---|---|
| 1 | 300 | medium | 10 | 150 | 2 |
| 2 | 600 | medium | 15 | 180 | 2 |
| 3 | 1200 | medium | 15 | 200 | 2 |

❑ After the heating step is completed, allow the contents of the reaction vessel to cool to 50°C or below before removing it from the microwave cavity.
❑ Carefully open the vessel in a fume hood to allow noxious fumes to escape.

## Prepare the Samples for Analysis

❑ Quantitatively transfer the digestate into a 25-mL volumetric flask and dilute to the mark with deionized water. This solution will be used for the analysis of iron and manganese.
❑ Dilute 1.0 mL of the above solution to 100 mL with deionized water. This 1:100 dilution is used for the analysis of sodium, potassium, calcium, and magnesium.

## Perform the Analysis

❑ Analyze the samples using atomic absorption, atomic emission, or inductively coupled plasma spectrometry.

## QUESTIONS

1. What is the purpose of the digestion step in the analysis of metals in dietary supplements?
2. The alkali metals, sodium and potassium, can be analyzed using atomic emission spectroscopy, whereas the alkaline earth metals, magnesium and calcium, and the transition metals, iron and manganese, must be analyzed using atomic absorption spectroscopy.
   a. What is the difference between atomic emission and atomic absorption?
   b. Why is atomic emission spectroscopy used for the alkali metals, but not for the alkaline earth and transition metals?
3. Dietary supplements are not regulated by the Food and Drug Administration and the manufacturer is responsible for assuring their safety. Do you think these supplements should be regulated by the FDA? Why or why not?

# Index

## S



## T



## U



## V



## W



## Z